T0325091

DANCE TO THE TUNE OF LIFE
BIOLOGICAL RELATIVITY

In this thought-provoking book, Denis Noble formulates the theory of biological relativity, emphasising that living organisms operate at multiple levels of complexity and must therefore be analysed from a multi-scale, relativistic perspective. Noble explains that all biological processes operate by means of molecular, cellular and organismal networks. The interactive nature of these fundamental processes is at the core of biological relativity and, as such, challenges simplified molecular reductionism. Noble shows that such an integrative view emerges as the necessary consequence of the rigorous application of mathematics to biology. Drawing on his pioneering work in the mathematical physics of biology, he shows that what emerges is a deeply humane picture of the role of the organism in constraining its chemistry, including its genes, to serve the organism as a whole, especially in the interaction with its social environment. This humanistic, holistic approach challenges the common gene-centred view held by many in modern biology and culture.

≈

Denis Noble is Emeritus Professor of Cardiovascular Physiology and Director of Computational Physiology at the University of Oxford, UK. He is the current President of the International Union of Physiological Sciences and a Fellow of the Royal Society.

DANCE TO THE
TUNE OF LIFE
BIOLOGICAL RELATIVITY

∼

Denis Noble
The University of Oxford, UK

CAMBRIDGE
UNIVERSITY PRESS

CAMBRIDGE
UNIVERSITY PRESS

University Printing House, Cambridge CB2 8BS, United Kingdom

One Liberty Plaza, 20th Floor, New York, NY 10006, USA

477 Williamstown Road, Port Melbourne, VIC 3207, Australia

314-321, 3rd Floor, Plot 3, Splendor Forum, Jasola District Centre, New Delhi - 110025, India

79 Anson Road, #06-04/06, Singapore 079906

Cambridge University Press is part of the University of Cambridge.

It furthers the University's mission by disseminating knowledge in the pursuit of education, learning and research at the highest international levels of excellence.

www.cambridge.org
Information on this title: www.cambridge.org/9781107176249

© Cambridge University Press 2017

First published 2017

A catalogue record for this publication is available from the British Library

Library of Congress Cataloging in Publication data
NAMES: Noble, Denis, 1936– author.
TITLE: Dance to the tune of life : biological relativity / Denis Noble.
DESCRIPTION: Cambridge ; New York : Cambridge University Press, [2017] | Includes bibliographical references.
IDENTIFIERS: LCCN 2016023673 | ISBN 9781107176249 (hardback)
SUBJECTS: lcsh: Life – Origin. | Relativity. | Cosmology. | Evolution (Biology) | MESH: Biological Evolution | Biodiversity | Genetic Variation | Origin of Life
CLASSIFICATION: LCC QH325.N625 2017 | NLM QH 366.2 | DDC 577 – dc23
LC record available at https://lccn.loc.gov/2016023673

ISBN 978-1-107-17624-9 Hardback

Contents

A colour plate section can be found between pages 174 and 175.

Preface

The central message of this book is that living organisms are open systems. That refers to all parts of organisms. All the molecules, organs and systems dance to the tune of the organism and its social context. Those molecules include the sequences of DNA we now call genes.

- How do all these components of life dance together in harmony?
- When did their billion-year dance begin?
- What makes them dance?
- Why is their dance relativistic?
- What do we mean by a 'gene'?
- What do we mean by 'life'?
- How can 'life' depend on 'dead' molecules?
- And what is Biological Relativity?

The answers to these questions form the subject of this book. We will also address the question of meaning. Could all this really happen as a consequence of 'blind chance'? And what could that commonly used phrase possibly mean? What, indeed, do we mean by 'meaning'? Could meaning itself be subject to a relativity principle: a relativity of epistemology?

If these questions fascinate you, then read on.

You will not need to know a lot of science to understand the book: what you will need is a new set of eyes. I will encourage the reader to adopt the eyes and mind of an inquisitive explorer. The scientific knowledge you need to know will mostly be in the book. If you already know a lot of science, you may need to relearn what you thought you knew. Because the central message is that twentieth-century biology went up the wrong street in the interpretation and presentation of its many impressive discoveries.

The reason is that some very influential twentieth-century biologists presented a simplistic gene-centred view of biology using memorable metaphors and brilliant writing to encourage you to adopt their view. And in this they were very successful. Hardly any biological discovery today is presented in the popular media without reference to the discovery of this or that gene 'for' something or other.

This book will show you that there are no genes 'for' anything. Living organisms have functions which use genes to make the molecules they need. Genes are used. They are not active causes.

This book will show you that there is no complete programme in our DNA. Programmes, if useful at all as a concept in biology, are distributed across scales in the organism.

This book will show you that there is no privileged level of causation, which is a central statement of the theory of Biological Relativity.

It will also show you that we are now far from certain what a gene is, and that many of the confusions and misrepresentations of biology arise from mixing up different definitions of genes and genetics.

We don't know when DNA first evolved. But it is virtually certain that it already existed two billion years ago. It seems likely that it must have existed for at least a billion years before that. There are fossils of the simplest cells that go back to over three billion years ago.[1] So, if genes dance, then they have been doing so for billions of years, in fact for most of the period of the Earth's existence, which is about 4.5 billion years.

For the Fainthearted

In spite of the sub-title of this book, don't be afraid if you are not mathematically trained. I promise you that, with the sole exception of Einstein's iconic equation $e = mc^2$, there are absolutely no equations in the main body of the book. Science could not function properly without mathematics. But, even in the most mathematical areas of science, and biology is rapidly becoming one of those, it is usually possible to explain the concepts in common language, once they have been distilled down from the abstract world of equations.

To help you through some uncharted territory, like the Bellman in Lewis Carroll's nonsense poem *The Hunting of the Snark*, remember that 'what I tell you three times is true'. I have deliberately included a certain amount of repetition in the different chapters, usually by expressing the same concept from a different angle or in a different context. Don't be alarmed if you think you have read something before. I turn some basic ideas in biology upside down. That takes a certain amount of getting used to. As you read on you may come to welcome those nice reminders of a point that is already half-appreciated. We are all used to this phenomenon in other ways. When we first see an unfamiliar object we easily mistake it for something else, and have to look again. That is even more true for unfamiliar concepts.

As an example, the fact that organisms are what we call open systems is employed in several chapters, from different perspectives. It is by appreciating the full extent of the development of this concept that a reader can come to understand its profound significance.

Although this book is critical of the simplistic way in which twentieth-century biology was often presented, my purpose is certainly not to minimise the phenomenal experimental achievements. It is rather an appeal for scientific humility. We are all prisoners of the cultures in which we find ourselves. Particularly in its theoretical aspects, science cannot be immune from culture even though it often challenges common and received ideas. Perhaps the ultimate principle of relativity is the relativity of knowledge, of epistemology. That is the title of the last chapter. As you journey from chapter to chapter, fasten your intellectual seatbelts. The ride through the book may jolt many of your present assumptions about the nature of living organisms.

The Sub-Title of the Book: A Challenge for the Future

The first complete draft of this book was finished in 2015, the centenary year of Einstein's General Theory of Relativity. That was not the initial reason for the sub-title, but it is a nice and appropriate coincidence. But, before the reader should judge me for being so presumptuous, let me hasten to add that what is developed in this book is more like a sketch when compared to the beautiful mathematical expressions of Special and General Relativity. Furthermore, I very much doubt whether the principle of Biological Relativity could be so expressed. We may not have the appropriate mathematics for an evolutionary process that has been as much a history as a phenomenon that could be predicted mathematically, except over relatively short time scales. Many biologists follow the lead of Stephen J. Gould in thinking that if the evolutionary clock could be set back to any point in history, the process would not follow the path that it has.

The extension of the principle of relativity to biology, as outlined here, is therefore more a set of signposts to a path. It opens up vistas that others better equipped than I might follow wherever they may lead. This is a challenge to younger scientists. I wrote the book while having the privilege of being the President of the International Union of Physiological Sciences. I believe it could be the union of those sciences with the relevant branches of physics, engineering and mathematics that could lead the way forward in the future.

Chapter Guide

Chapter 1 introduces the general principle of relativity as it developed in the study of the universe. Understanding the steps by which the idea of relativity was reached will prepare you for application of the general principle to biology, which is the core of the book.

Chapters 2–4 contain the background knowledge of biology required to understand the later chapters. Chapter 2 is a complement to Chapter 1 since instead of reaching out to the larger scales of the universe as a whole it reaches down to the microscopic and molecular components of our bodies. It will guide you through the various levels of organisation from molecules to the whole organism. Chapter 3 then introduces the processes that characterise life in the form of networks of interactions. I will give some examples of networks that involve multiple levels. Multilevel interactions form a central aspect of Biological Relativity since causation is then not restricted to one level and is necessarily bi-directional. Chapter 4 shows how these components and processes work in the smallest living things – single cells. The great majority of organisms on Earth are unicellular, and even multicellular organisms go through a single-cell stage when they reproduce.

Chapter 5 outlines the current widely held theory of evolution (Neo-Darwinism) and analyses its main conceptual problems. You will learn that it is a gene-centric, molecular-oriented view of biology. By focusing on genes and molecules it cannot answer the question 'what is life?' Moreover, it was not Darwin's theory of evolution.

Chapter 6 explains the central principle of Biological Relativity. You will learn that organisms are alive precisely because their processes operate at and between many different scales and levels. The molecular and other components are constrained by all levels, including the environment.

Chapters 7 and 8 describe the experimental findings that enable an integrative relativistic theory of evolution to be developed to replace Neo-Darwinism. Chapter 7 focuses on the ways in which the genetic material, DNA, has been rearranged during evolution. Chapter 8 focuses on the epigenetic and related mechanisms by which the genome is controlled.

Chapter 9 returns to the questions asked in Chapter 1 and develops a form of relativity of our knowledge of the universe: a relativity of epistemology. It is through this idea that we arrive at answers that science can give to the big questions about the universe and ourselves and to an understanding of the limits of those answers.

Chapter 10 is written as a brief postscript that summarises the central argument of the book.

Each chapter begins with an easy way in, often using stories from my personal experience. As you read on, you will see the relevance of the story to the main message of the chapter.

You might initially wonder how such a diverse range of topics hangs together since the book begins with the fundamentals of physics and cosmology, yet ends with the fundamentals of biology and the limits to our knowledge. You will discover, perhaps surprisingly, that there are many links between these various threads. The insights of Chapter 1 inform important conclusions in many of the subsequent chapters, and the general principle of relativity informs the whole book.

It will be clear from this introduction to the various chapters, and how they link together, that this book is not a textbook of the systems approach to biology. My aim is rather different. It is to contribute to the new trends in biology that have become evident during the first decade or so of the twenty-first century by creating a coherent conceptual framework within which those trends and their experimental basis can be understood. In any case, there is no need for me to write a textbook since an excellent one has been published already: Capra and Luisi's (2014) *The Systems View of Life: A Unifying Vision* (Cambridge University Press, 2014). At various points in my book I will cross-reference this text to guide readers to the relevant parts of their book. Their vision of the systems approach is very similar to mine.

Notes and glossary. The glossary is an important part of the book. Some key words have significantly different interpretations and definitions used by different writers. These include reductionism, Neo-Darwinism, Darwinism, Lamarckism and epigenetics. When you first encounter these words, you may benefit from consulting the glossary entries on them.

Note

1 Fossils of microbes metabolising sulphur have been identified in rocks dating from 3.4 billion years ago: Wacey, D., M.R. Kilburn, M. Saunders, J. Cliff and M.D. Brasier (2011) Microfossils of sulphur-metabolizing cells in 3.4 billion-year-old rocks of Western Australia. *Nature Geoscience* 4:698–702.

Acknowledgements

Working towards writing this book has been a very long journey. Many colleagues, friends and critics have been companions on that journey.

First, both in time and in what I owe him, is my PhD supervisor, Otto Hutter, who first set me out on the journey at University College London way back in 1958. Even nearly 60 years later he is still my best critic, and kindly read many of the draft chapters.

Second is my brother, Ray, who has inspired many of the ideas of this book ever since his undergraduate days in zoology at Manchester University and more recently as a medical ethicist at University College London. He spotted the problems with gene centric accounts of biology well before I did.

Third are the innumerable students who have studied with me and, in the process, often taught me their own wisdom over a period of 50 years. Nothing can prepare you for the 'wow' moment when a student brings a razor sharp new mind to an old problem and cuts through the standard textbook guff.

Fourth are fellow academics from all over the world who have criticised and helped to smooth the wilder aspects of my journey. They have particularly included scientists and philosophers at Balliol College over many years. I am deeply privileged to have worked in such a richly interdisciplinary Oxford college.[1] Some of the lectures and videos referred to in this book were recorded by *Voices from Oxford*, based at Balliol College, and I am very grateful to the Director, Professor SungHee Kim, for all the advice and help she and the *Voices from Oxford* team have given.

Finally, I especially thank those who have trenchantly disagreed with me. Some of them may well say that I didn't take much notice of them. Not really true. But it is true that often enough they influenced me in ways that they might not recognise.

An intellectual journey in which you end up in a place very different from your starting point can often be lonely, a kind of pilgrim's progress with many doubts on the way. To all who have helped, hindered or just lent a kindly ear, I thank you.

The full technical details for parts of this book were first published as invited articles in *Science, Molecular Systems Biology, Philosophical*

Transactions of the Royal Society, Interface Focus, Journal of Experimental Biology, Journal of Physiology, Experimental Physiology, Physiology News and other journals and books published between 2008 and 2015. I am grateful to the editors of these journals and books and to the referees for many valuable comments and criticisms. The ideas in this book have been through extensive peer review.[2]

Finally, I thank those who kindly read and criticised early drafts of particular chapters or of the whole book. I am particularly grateful to Geoffrey Bamford, Sir Patrick Bateson, Nicholas Beale, Steven Bergman, Dario DiFrancesco, Yung Earm, Martin Fink, Otto Hutter, Eva Jablonka, Mike Joyner, Sir Anthony Kenny, Ard Louis, Colin Meyer, Derek Moulton, Raymond Noble, Susan Noble, Kazuyo Tasaki, Toshiaki Tasaki, David Vines and Michael Yudkin, many of whom gave me valuable feedback and detailed corrections. I also thank the Press reviewers for very helpful feedback.

Notes

1   The philosophers include particularly Stuart Hampshire, Charles Taylor, Alan Montefiore, Anthony Kenny and Peter Hacker.
2   The full list of these publications for those who want to study the technical detail is as follows, with the key publications starred:

Systems biology:

* Noble, D. (2008) Claude Bernard, the first Systems Biologist, and the future of Physiology. *Experimental Physiology* 93:16–26.

Auffray, C. and D. Noble (2009) Conceptual and experimental origins of integrative systems biology in William Harvey's masterpiece on the movement of the heart and the blood in animals. *International Journal of Molecular Sciences* 10:1658–1669.

Kohl, P. and D. Noble (2009) Systems biology and the virtual physiological human. *Molecular Systems Biology* 5:291–296.

* Kohl, P., E. Crampin, T.A. Quinn and D. Noble (2010) Systems biology: an approach. *Clinical Pharmacology and Therapeutics* 88:25–33.

Noble, D. (2010) Mind over molecule: systems biology for neuroscience and psychiatry. In *Systems Biology in Psychiatric Research*. F. Tretter, G. Winterer, J. Gebicke-Haerter and E.R. Mendoza, editors (Wiley-VCH, Weinheim; pp. 97–109).

Noble, D. (2010) Biophysics and systems biology. *Philosophical Transactions of the Royal Society A* 368:1125–1139.

Noble, D., R. Noble and J. Schwaber (2014) What is it to be conscious? In *The Claustrum*. J. Smythies, V.S. Ramachandran and L. Edelstein, editors (Academic Press, San Diego, CA; pp. 353–364).

Genes and causation:

* Noble, D. (2008) Genes and causation. *Philosophical Transactions of the Royal Society A* 366:3001–3015.

Noble, D. (2008) For a redefinition of god. *Science* 320:1590–1591.

* Noble, D. (2011) Editorial. *Interface Focus* 1:1–2.

Noble, D. (2011) Differential and integral views of genetics in computational systems biology. *Interface Focus* 1:7–15.

Ellis, G.F.R., D. Noble and T. O'Connor (2012) Top-down causation: an integrating theme within and across the sciences. *Interface Focus* 2:1–3.

* Noble, D. (2012) A theory of biological relativity: no privileged level of causation. *Interface Focus* 2:55–64.

Noble, D. (2013) A biological relativity view of the relationships between genomes and phenotypes. *Progress in Biophysics and Molecular Biology* 111:59–65.

Evolution:

Noble, D. (2010) Letter from Lamarck. *Physiology News* 78:31.

Noble, D. (2011) Book review: Evolution. A view from the 21st century. *Physiology News* 85:40–41.

* Noble, D. (2011) Neo-Darwinism, the modern synthesis, and selfish genes: are they of use in physiology? *Journal of Physiology* 589:1007–1015.

Noble, D. (2013) Life changes itself via genetic engineering. Comment on 'How Life Changes Itself: The Read-Write (RW) Genome' by James Shapiro. *Physics of Life Reviews* 10:344–346.

* Noble, D. (2013) Physiology is rocking the foundations of evolutionary biology. *Experimental Physiology* 98:1235–1243.

Noble, D. (2014) Secrets of life from beyond the grave. *Physiology News* 97:34–35.

* Noble, D., E. Jablonka, M.M. Joyner, G.B. Müller and S.W. Omholt (2014) Evolution evolves: physiology returns to centre stage. *Journal of Physiology* 592:2237–2244.

* Noble, D. (2015) Evolution beyond neo-Darwinism. *Journal of Experimental Biology* 218:7–13.

Noble, D. (2015) Conrad Waddington and the origin of epigenetics. *Journal of Experimental Biology,* 218:816–818.

Noble, D. (2015) Central tenets of neo-Darwinism broken: response to 'Neo-Darwinism is just fine'. *Journal of Experimental Biology* 218:2659.

1

The Universe and the Principle of Relativity

'Let there be light, and there was light'
יהי אור ויהי אור
(Genesis 1:3) (Figure 1.1)

The Sky at Night

Now that most of humanity lives in cities, it has become rare for us to experience the full extent of the wonder that our predecessors must have felt as they saw the night sky from open country or from their unlit dwelling places. On every clear moonless night they would have experienced what we can do only by going to remote parts of the countryside far away from the city lights. They would have noticed that as dusk gives way to night, more and more stars appear. And as their eyes slowly adapted to the dark, even more would appear until they became uncountable. On a really clear, cold night they would also experience the feeling that the universe is somehow alive with activity as the faintest stars seem to appear and disappear depending on whether one looks directly at them or at a slight angle. There is a depth also to the sheer blackness of space between the stars that contrasts so markedly with its light blue during the day. The sky at night viewed in this way, when there is little or no moonlight, is a miracle, with a giant belt (the Milky Way) running across it, with countless stars appearing, more the longer we look, and with the occasional larger movement as a meteor appears. To crown the spectacle, it moves slowly and majestically throughout the night.

1

Figure 1.1 A view of the Milky Way towards the Constellation Sagittarius (including the Galactic Centre) as seen from a non-light polluted area (the Black Rock Desert, Nevada) (courtesy of Steve Jurvetson). For a colour version of this figure, please see the plate section.

Faced with such glory and spectacular beauty, we are forced to ask a question. Why?

The question pushes its way before us. And the human response to this question has always been the same: to propose an answer. We find it difficult to live without answers. That is what drives our metaphysical instincts, which in turn create our systems of religious and scientific thought. They are not so far apart as many might think. The quest for meaning can be seen as the religious instinct. The quest for explanation in terms of cause can be seen as the scientific instinct. But the two connect through the fact that we cannot even begin to develop an explanation without making some meaningful assumptions about the framework within which we can interpret what we see, feel and hear. We need a metaphysics within which we can develop our physics. That is as true today as it was in the earliest scientific discoveries, as we will see as the story in this book develops. Science also contributes to understanding meaning through identifying what we call function. It is too simplistic to say that science deals only with 'how?', while religion deals only with 'why?'. The two questions intertwine.

So, what did our ancestors do to make sense of what they saw in the darkest of nights in the deep countryside? They saw groups of stars, what we today call constellations. They also imagined that these groups had meaning and so they gave them names. There is one particular constellation, the one we now call the Big Dipper or The Plough, which received names in all the main historical traditions we know about. Many saw a bear, which is why its Latin name is *Ursa Major*, the great bear. It appears in Babylonian and Egyptian astronomy, leading to the Greek system, and in the Jewish system which leads to its reference in the Bible.[1] It appears in ancient Chinese[2] and Indian[3] astronomy, and in every other traditional system. In fact, this constellation is only one of two (the other is *Orion*) that appear as such in both the Western and Eastern astronomical traditions. The Chinese divided the sky up in a different way, based on the pole star, Polaris, whereas the Greeks thought in terms of the relationship of the constellations to the way the sun appears to move amongst them during the year, which is what gave us the signs of the zodiac.

Dividing up the sky into constellations was very practical. Relating them to the pole star was particularly helpful to travellers and mariners. *Ursa Major* points towards the pole star and could therefore be used to find north. It was also possible to use the sky as a timekeeper since it rotates smoothly throughout the night. If one knows the constellations well and how their movements change during the year, one can work out what time of night it is. All of this was important to people navigating through open seas and deserts. The sky was their signpost and clock. Those highly practical results arose from the smooth circular movement of the heaven above us as it rotates around the Earth.

Or does it? Today, we know that assumption is wrong. But it is instructive to understand the steps by which we came to that conclusion. Therein lies the origin of the principle of relativity.[4] Most people think that the principle applies only to physics. One of the purposes of this book is to show that, in its widest sense, it must apply also to biology. First we must understand its use in physics, which is the purpose of this chapter. We will then be able to explore its impact on biology.

Before I outline the steps by which the fundamental principle of relativity developed, I would like to ask the reader to adopt the attitude of an inquiring explorer. It is easy for us to laugh at what we see as the misunderstandings of the past. A flat earth? Absurd! A heavenly globe containing the stars? Ridiculous! With that attitude, it is also easy to forget that we will be seen as ancestors in the future. How do we know that we, and we alone amongst the tens of thousands of years of human thought, have at last got the answers right? Many thoughtful scientists today are convinced that there are more revolutions to come and are not at all happy with our current models of nature.

Those models are brilliantly successful at prediction, much more so than ever before. But as a basis for understanding, for feeling certain that we have 'got it', they leave much to be desired. We find it difficult, for example, to unify the physics of the smallest scales, where quantum mechanics is relevant, and the physics of large scales, where general relativity dominates. Nor do we know how to explain the apparently arbitrary nature of the constants of the universe,[5] although we know that they need to be within narrow limits for our universe to exist and for living systems to be possible. In biology, there are many more puzzles calling out for answers: what is life? How did we as a species get to be the way we are? What is a gene? And many more. In the search for those answers, we followed a largely blind alley during the twentieth century. The blind alley is the idea that the genome is the 'book of life', a blueprint from which you and me, and all other living creatures, were made.

We have more to learn from the history of thought about the universe than we might think. If we take each step seriously and understand why it was taken, we will then understand better what steps we can take in the future to distance ourselves from our own misunderstandings. This book is also an appeal for humility in scientific thought. It occupies an intellectual space billions of miles from the naive certainties that many popular science writers portray. We advance in understanding by first coming to know what we don't understand. That kind of knowledge requires hard work. We have to undo some of our cherished basic beliefs.

So, join me on a thoughtful and provocative journey through the questions that we can't help asking. We begin in this chapter by asking how to interpret the sky at night, how that question led to the principle of relativity, and to the Special Relativity and General Relativity forms of the theory proposed by Einstein.[6]

Early Cosmologies

The oldest Hebrew sources represented the Earth as a flat disc floating in a huge sea. Since no one could consider the possibility of going completely round the Earth, the idea that the habitable Earth must have an edge, beyond which was a sea, was a reasonable assumption. The heavens were then represented as a hollow sphere with the stars set in the surface of the sphere as points of light in what could be viewed as a massive celestial candelabrum. Clearly the sphere must move, which creates the difficult question of where it goes when it moves below the horizon. And there must be several such spheres since the sun moves separately, and so do the 'stars' that we now know are planets.

One way to think about such a universe is that, since it consists of concentric spheres, perhaps its centre is also a sphere. That makes it easier to answer the question of where the spheres go when they disappear below the horizon. They just go round the central sphere, which must be the Earth. We don't know when exactly the idea that the Earth too was a sphere first arose, but we do know that it was a central idea for the astronomer Claudius Ptolemy, who lived around CE 90 to about CE 168. As his given name, Claudius, suggests, he was a Roman citizen, although he lived in Egypt when it was ruled from Rome, and his family name, Ptolemy, is Greek. He wrote in classical Greek.

He is said to have used Babylonian astronomical data to construct an elaborate set of tables and mathematical calculations brought together in the first surviving textbook of astronomy, called the *Almagest*. It includes ingenious geometrical calculations from a Greek mathematician, Hipparchus, which allowed estimations of the distances from the Earth to the sun and the moon. These calculations enabled the celestial spheres to be given dimensions and distances. In addition to the sphere carrying the sun, additional spheres carried the planets, and of course the outermost sphere carried the stars. In addition to the Earth, there were eight spheres carrying the sun, the moon, five known planets (Mercury, Venus, Mars, Jupiter, Saturn) and the fixed stars.

This shift in perception about the Earth and the universe can be represented as the first stage in developing the principle of relativity. As I will use this principle in this book it consists of distancing ourselves from privileged viewpoints for which there is insufficient justification. There are no absolutes – rather, even in science things can only be understood in a relative sense: relative to the question we ask; relative to the scale at which we ask the question; relative to our present knowledge of a universe of which we will always have questions remaining. In this sense, a privileged position is akin to an absolute.

Coming to view the Earth as yet another sphere was precisely such a use of relativity. The Earth was no longer viewed as a uniquely flat object.[7] Like the rest of the universe, it became a sphere. You will learn how this very general principle of distancing ourselves from supposedly unique or privileged viewpoints leads to the more familiar theories of relativity later in this chapter, and then to the theory of Biological Relativity in Chapter 6.

Distancing ourselves from viewing the Earth as a flat object may not have been easy. Many nineteenth-century writers thought that the idea of a flat Earth was originally so convincing that when Christopher Columbus set off in 1492 to sail west in order to arrive at the east, uneducated people still feared that he might reach the edge of the Earth, and perhaps never be seen again. This is a modern myth.[8] Medieval scholars were quite clear that the Earth was round. The mistake Columbus made was to calculate that East Asia was much closer. Finding the Caribbean islands saved him and his crew, and he still believed he had found the East.

The Copernican Revolution

Distancing ourselves from the geocentric view takes us to another great astronomer, Nicolaus Copernicus. Copernicus was born in 1473 in the Kingdom of Poland, and studied first at the Jagiellonian University in the then capital, Krakow. Later he moved to Italy and the famous universities of Bologna, Padua and Ferrara. Along with the anatomist Vesalius, he became one of the great polymaths of the Renaissance and a trigger of the Scientific Revolution. His book on the universe, *De revolutionibus orbium coelestium* (on the revolutions of the heavenly spheres) was published just before his death in 1543.[9]

The idea that the sun might be the centre of the universe was not entirely new. Similar ideas had been proposed in the third century BC by Aristarchus. We know this because Archimedes describes how Aristarchus thought that the fixed stars and the sun are unmoved, while the Earth revolves around the sun. Aristarchus also correctly thought that the fixed stars were very far away.

Copernicus, though, deserves the credit for providing the mathematical basis to show that the idea was predictive, and that it explained the strange fact that on the geocentric view the planets seem sometimes to move backwards. The Ptolemaic system had been made more complex in order to deal with this problem by postulating the existence of further epicycles.[10] This illustrates a pattern that is often repeated in science. Well-loved theories do not usually die suddenly. People try to find ways of retaining the central ideas of the theory while adding complexity to the explanations to accommodate observations that do not seem to fit the theory. Sometimes, the adoption of a new theory depends more on the overall weight of evidence, rather than on a single knock-down observation. We will have the opportunity to observe this process in science in later chapters of this book.

The clarity with which Copernicus stated the heliocentric view was impressive. He expressed his ideas as seven assumptions. First, that there is not one centre. This was to allow him to retain one aspect of geocentrism, which is that the moon orbits the Earth. This was explicitly stated in his second assumption, together with the statement that the Earth is not the centre of the universe. Instead, in assumption three he stated that the sun is or is near the centre of the universe. He then made a remarkable deduction. From observations of distances, he concluded that the distance to the stars is much greater than the distance of the Earth from the sun. His fifth assumption is also remarkable. This is that the motion of the stars represents simply the spinning motion of the Earth. The fixed stars are just that: fixed, immoveable. Assumption six gives the Earth an additional motion, that of orbiting the sun. Finally, he explained the Ptolemaic apparently 'backwards' movements of the planets as due to the Earth's motion.

This was the second application of the general principle of relativity. Arguably, it may have been the most important one since abandoning the privileged position of the Earth as the centre of the universe was a first step. That idea leads inevitably to the more familiar theories of relativity since, once we abandon the idea that our home, the Earth, is in any way special, why should we be convinced that anything else is the centre of the universe?

But it did not do so immediately. In fact, the ideas of Copernicus did not initially create any great waves of controversy. Significantly, there was no dramatic argument with religious thinkers. This fact is very important in the story of this book. Religious thinkers treated Copernicus as they treated themselves: as metaphysical theorists struggling to make sense of the world and the universe. This is hardly surprising since he was also a canon of the Catholic Church. Moreover, other church leaders had also proposed ideas similar to those of Copernicus. Two centuries earlier the French bishop Nicole Oresme had considered the proposition that the Earth rotates. In fact, his *Livre du ciel et du monde* contains the spirit of relativity, since he showed that to assume that the Earth is rotating rather than the heavens would not change any of the astronomical calculations. He appreciated the fact that different metaphysical standpoints can lead to the same conclusions concerning relative movements. His work did not, however, lead to a revolution of thought in the way that Copernicus' work did. In fact, he concluded that there was no proof that the Earth rotates – and no disproof either!

In the fifteenth century, the German Cardinal Nicholas of Cusa had expressed in a book called *De Docta Ignorantia* (roughly translated as 'On scientific ignorance') a viewpoint that is infused with the central idea of relativity: 'Thus the fabric of the world will have its centre everywhere and circumference nowhere.' This is remarkable since it also anticipates the later stage of questioning whether even the sun (or any other point)

Figure 1.2 Jupiter and the four Galilean moons observed through a Meade "10" LX200 telescope, i.e. ten times more magnification than was available to Galileo (Jan Sandberg, Wikimedia). For a colour version of this figure, please see the plate section.

could be the centre of the universe. Not surprisingly, he also developed a sophisticated, some would say mystical, concept of god.[11]

These historical facts are important. They show that the widely held view that every major advance in science has provoked reaction from conservative religious thinkers is far too simplistic. The more accurate historical view is that these debates about the nature of the universe occurred as much within the Church as outside it. Arguably, Nicholas of Cusa was the greater revolutionary than Nicolaus Copernicus since he was way ahead in questioning even the idea of giving a privileged position to the sun, *or any other celestial object.*

As to opposition to Copernicus, there were opponents both within and without the Church. Wider scientific acceptance of his ideas had to wait for more experimental proof anyway. This came with the work of Galileo and the first use of the telescope (Figure 1.2).

Galileo: Father of Modern Science

Galileo Galilei was born in 1564 and studied medicine at the University of Pisa. It was Einstein who called him the 'father of modern science'. He transformed our study of the universe. He did so using his own early telescope of very limited power (magnification about $\times 20$), so with even a modest modern telescope you can easily repeat some of Galileo's key observations, which he made on 7 January 1610.

The planet Jupiter can often be seen as a bright object. Amongst the planets, only Venus is brighter. Its position in the sky depends of course on its movements, so you need to consult a guide to its position on any given night. It is easily the largest planet, a gas giant 11 times the Earth's diameter. Unless there can be living systems very different from what we know, it could not support life. However, it has many moons and four of these are so large that they could be observed by Galileo. You can also see them. They are arranged on the same plane so you will see them strung out on either side of Jupiter. They orbit Jupiter in a matter of days, so you can also repeat another of Galileo's observations, which is to see that they are in different positions every night. Galileo, of course, saw the point. Here is a miniature solar system with Jupiter acting as the attraction in place of the sun and the moons playing the role of the planets. It is hard to make these observations without realising that the Earth must also go round the sun. And that the planets that do so can have moons just as the Earth has a moon. While Jupiter itself is very unlikely to harbour life, its moons might do so. Europa has a surface of ice and water which might well support life.

Galileo's observations and his defence of the heliocentric idea came about 60 years after Copernicus' publication of his work. This time, the mood within the Church was different. Some, notably amongst the Jesuits, supported him. But it is thought that intrigue at the Vatican led to Urban VIII, who had been a supportive friend, even encouraging him to publish his work, becoming offended by what could be seen to be mockery of him and the geocentric view in Galileo's book *Dialogue Concerning the Two Chief World Systems*.[12] The defender of the geocentric view was a character called Simplicio, which carries the connotation of simpleton. Offending friends by mocking them may not be wise. Perhaps Galileo meant no offence. Simplicio was simply a literary device.

There have been many books and articles written on these events and the subsequent famous 'recantation' of Galileo.[13] It is true that Galileo was found guilty of heresy by the inquisition and put under house arrest, while his books were banned. The ban on his books was not lifted until the eighteenth century. Famously, in 1992 Pope John Paul II expressed regret for the events that led to the Church accusing him of heresy and subjecting him to house arrest.

It is right to condemn the seventeenth-century Vatican inquisitors. They were certain they were right and Galileo was wrong, so wrong that

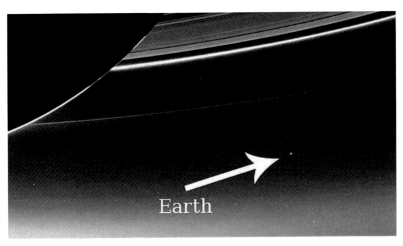

Figure 1.3 The Earth (little blue dot) viewed from the spacecraft *Cassini* as it photographed the rings of Saturn during an eclipse of the sun by Saturn. *Cassini* was 900 million miles from Earth. Light takes over an hour to reach the Earth from Saturn. But this is minuscule compared to the more than 13 billion years for light to reach us from the edge of the observable universe (source: NASA/JPL-Caltech/Space Science Institute (www.nasa.gov/mission_pages/cassini/multimedia/pia17171.html)). For a colour version of this figure, please see the plate section.

he had to be humiliated and punished. It took nearly four centuries for the injustice to be openly acknowledged. It is one of the strange characteristics of metaphysics that, while its very speculative nature should convince people to be cautious, even humble, it often does just the reverse. Perhaps the very uncertainty creates the inner wish for certainty. That, after all, is also part of the religious instinct – the search for the certainty of faith. Scientists, even atheistic scientists, are not immune to the same problem. If you think that could not happen, that scientists could not be so cruel, wait until you read in later chapters the way in which twentieth-century scientists ridiculed the great French biologist Jean-Baptiste Lamarck and sidelined almost completely the brilliant developmental biologist and polymath Conrad Waddington.

The Earth from a Billion Miles (Figure 1.3)

Galileo would surely have been delighted to see the Earth from the giant planet Saturn, as we can now do thanks to the voyage of the spacecraft

Cassini. He would have seen the Earth as a tiny blue dot just as we view other planets as small objects in the solar system. Hardly a candidate for the centre of the universe! Outside our solar system the Earth would not even be directly visible. Its presence could be detected only indirectly, just as we now detect what are called exoplanets, planets circling other stars to produce a tiny fluctuation in the light from them, and the tiny perturbations of the star's position.

Newton's Laws of Motion

It is one thing to show that planets orbit the sun, and moons orbit the planets. It is quite another thing to show how those motions could be explained. Our own experience teaches us that for something to move continually there must be a force making it do so. Mechanics before Newton adopted the same common sense view. Without a force, an object would stop moving. Newton reversed that. Without a force it would continue moving! Not just temporarily, like the supermarket trolley after a brief push, but *permanently*. It would never stop. Imagine the trolley continuing on its motion until it goes right around the Earth and returns to you! A satellite in orbit does precisely that.

Newton's laws are so familiar to us now that it is difficult to imagine how counterintuitive they must have seemed in his day. People were so used to the idea that hard work had to be done to move anything around, whether on a farm, in the house or along the streets. If the planets moved indefinitely then something (angels?) had to be moving them.

Newton was born in 1643 and had quite a difficult childhood. He studied at Trinity College Cambridge and graduated in 1665 just before the university closed because of the plague. It was while at home for two years that he developed his brilliant mathematics. This included calculus, his optics and the law of gravitation. Not bad for a two-year stint at home! He returned to Cambridge in 1667 to become a Fellow, and only two years later was given the prestigious Lucasian Chair. He was unorthodox enough to avoid the rules that, at that time, required all the Fellows and Professors to be ordained. One can speculate that it was his very unorthodoxy that stood him in good stead in challenging so many ideas of his time.

Newton must have realised that, of course, there is an everyday example of continual motion: a falling object does not stop until it hits the ground. On the contrary, it accelerates. There was therefore no need to suppose that the moon needs something to push it round the Earth. It just goes on falling. But because it already has lateral motion it will combine that motion with the fall due to gravity to produce overall motion in an orbit. If the moon was in deep space it would just travel indefinitely in a straight line.

This insight established several important things that are relevant to the principle of relativity and to our story. The first is that celestial objects experience the same forces and motions as those on Earth. We prove that today every time we send a satellite into orbit. If we give the satellite enough speed it will 'escape' the Earth's gravitational attraction and go into orbit. Actually, it never escapes gravity. It is rather that the force of gravity and the inertial motion we have given it combine, just as they do for the moon, to enable it to 'fall' continually in an orbit around the Earth. The second is that if we know the forces acting on objects, we could use calculus to predict their motions indefinitely.

There was, however, one problem relating to the principle of relativity that Newton had difficulty solving. Where was the centre of the universe? Could it be the sun? But he also understood that the centre of gravity (which might be a possible interpretation of 'centre') of the solar system was not precisely the sun. He decided that 'the common centre of gravity of the Earth, the Sun and all the Planets is to be esteem'd the Centre of the World'. If he had gone one step further and recognised, as Nicholas de Cusa had, that 'the world will have its centre everywhere' he would have made the next step into the fully relativistic idea that there is no centre, and that all movement is relative.

Nineteenth-Century Certainties

With Newton's equations of motion and the idea that the universe, and perhaps everything in it, worked rather like clockwork, it seemed that in principle everything could be predicted with just the force of gravity and the laws of motion. The conviction that this must be so became very strong in the middle of the nineteenth century. Even the apparently

strange phenomena of electricity and magnetism could be accommo-
dated as Faraday and Maxwell showed how they could be used to produce
movement, and in turn be generated by movement.

The bible of this certainty was the book *Celestial Mechanics*, produced
by the French mathematician Pierre-Simon Laplace as a series of volumes at the beginning of the nineteenth century. He was such a brilliant
mathematician that many people called him the French Newton. In his
Philosophical Essay he wrote 'all the effects of nature are only mathematical results of a small number of immutable laws'. The clear impression
was that, in principle, everything could be known and accurately calcu-
lated in a deterministic universe.[14]

It is very relevant to an important message of this book that two
brilliant minds, Newton and Laplace, using essentially the same mathematical tools and facts about the universe, could come to diametrically
opposed metaphysical conclusions. Newton perceived no conflict at all
with his belief that 'God is Eternal and Infinite, Omnipotent and Omni-
scient. . . . He endures forever, and is everywhere present; and by existing
always and every where, he constitutes Duration and Space.'[15] Laplace,
living more than a century later, came to the opposite conclusion. He
may or may not have told Napoleon Bonaparte 'I have no need of that
[the God] hypothesis', but the quip certainly reflected his atheistic interpretation of Newton's mechanics.

Yet, according to another French mathematician, Joseph Fourier,
Laplace's last words were 'What we know is little, and what we are igno-
rant of is immense.' I like the idea that even Laplace was so cautious.

His influence on other nineteenth-century scientists was profound.
In biology, in particular, he clearly influenced the physiologist Claude
Bernard. This is very relevant to the story in subsequent chapters of this
book.

With the universe safely wrapped up in Laplace's interpretation
of Newton's equations of motion and with even the most innovative
thinkers in biology, like Claude Bernard, convinced that all they needed
to do was to use the understanding of the physical sciences in studying biology, it must have seemed that science was well on the path to its
ultimate goal in the explanation of everything. The great physicist Lord
Kelvin expressed this great confidence in 1900: 'There is nothing new to
be discovered in physics now. All that remains is more and more precise
measurement.'[16] The work of Darwin, Wallace and others on the theory

of evolution, to which we will return in subsequent chapters, reinforced that view since it promised an understanding of the living as well as of the inanimate world.

But waiting in the wings were two discoveries that were to shatter this confidence so far as physics is concerned. They were quantum mechanics and the further developments of the principle of relativity in the form of the special and general versions of relativity of Einstein's theories. Just five years after Lord Kelvin wrote those famous words, they had to be eaten.

Quantum Mechanics[17]

The vision of Laplace perfectly expressed nineteenth-century scientific optimism. It must have seemed that it was only a matter of time before we could determine where objects are in the world, how they are moving, and then set the calculators to work. We would then understand what it must be like for a god to 'see' everything at once. The equations of Newtonian mechanics work just as well going backward as forward. In principle, therefore, we would be able to know how the world was at all previous times, how it is now and how it will be at all future times. The dimension of time would indeed be just an additional, fourth, dimension like the three spatial ones.

But can we know where every object is, and how it is moving? Surely that can only be a matter of how good our instruments may be. And, however we interpret it, time does have a direction!

The experiments that led to the development of quantum mechanics show that it is not possible at the micro level to determine both the positions and velocities independently and with any degree of accuracy. The experiments show that particles behave as waves under some circumstances and as particles under other circumstances. They behave as particles when they hit a detecting instrument in a particular place and at a particular time. But they behave as waves when, for example, two of them are allowed to interact to produce the beautiful interference patterns that are characteristic of colliding waves. There is a particle–wave duality. To say the least, that is difficult to visualise. We can visualise a wave as the behaviour of a large number of elements, such as water molecules in a lake, where the wave is a transmitted behaviour of the ensemble.

No single water molecule travels with the wave. The molecules simply bob up and down, much like a Mexican wave in a sports stadium. Similarly, a sound wave is the transmitted behaviour of large numbers of molecules in the medium, air or fluid or solid, that transmits it. The molecules of the medium oscillate back and forth but they do not travel with the wave. But a particle is a discrete object which should be part of an ensemble to allow a wave to form. How can it also be the wave itself? The particle that is itself the wave travels with the wave. And when it is behaving as a wave, what is the medium in which the wave occurs? As we will see, that question connects with Einstein's Special Theory of Relativity, discussed in the next section.

A further difficulty is that we can only say where the particle might be with a certain degree of probability. The same applies to its velocity. There is a fundamental degree of uncertainty such that the more accurately we try to determine position, the less certain we can be about velocity, and vice versa.

There are many other ways in which the features and consequences of quantum mechanics can be expressed, but this characterisation will suffice for the purposes of this chapter.

Physicists and philosophers have thought deeply about the implications of quantum mechanics. Early reactions were that this can't be true, or at least only provisionally true while waiting for something to replace it. Einstein was very sceptical; he said: 'God does not play dice.' Yet, she does! There seems to be no way around the shocking nature of this discovery. It shakes the very foundations of nineteenth-century confidence. People have therefore tried various ways of arguing for minimising the impact. One of these is to say that this uncertainty applies only at the micro level, the subatomic level. That is not entirely true. There are conditions, such as very low temperatures, under which quantum mechanical behaviour has been shown to exist at a macro level. And people are already using quantum mechanical properties to construct macro-level machines.

Quantum computers are a good example. They use quantal behaviour to implement more logical operations simultaneously than can be done with conventional computers, and experiments have already been done to demonstrate the feasibility.[18] But of course the machine itself must be usable by a human being. The quantal properties at the micro level will have consequences for what happens at the macro level. That has already been shown by maintaining quantal memory states that could be used in such a computer at room temperature for more than 30 minutes.[19] The first demonstrations of such effects were at exceedingly low temperatures, beyond the range of living organisms. Some have also speculated, perhaps wildly at this stage, that there could be features of our brains that allow quantum mechanical effects to play a role.[20]

Furthermore, this way of dealing with the problem is not really satisfactory to someone who wants to know what the world is really like. The best way, at present at least, to interpret the equations of quantum mechanics is to note that they work. In fact they work very well for describing what we see. But they don't provide a satisfactory explanation of the world 'as it really is'.

This returns us to the big 'why?' question posed at the beginning of this chapter. Possibly, it is the wrong question. There may be aspects of reality that we can never know. That is not a comfortable position to be in. Our predecessors asking the big 'why?' question on looking at the sky at night would hardly find this kind of answer satisfactory.

Another approach to this problem is to say that there must be more to be found out that may lead to a set of physical theories that are more satisfactory. That is the approach of those who note that there are also other unsatisfactory features about our present knowledge, not least that we use different theories for micro and macro levels. Perhaps we should just wait until another Einstein turns up to sort it all out.

I will return to these questions in the last chapter of the book. Meanwhile, our story moves on to Albert Einstein.

Einstein's Special Theory of Relativity

A recurring problem with the various stages of application of the relativity principle has been the persistent idea that there must always be a medium in which movement occurs. Early objections to the view that the Earth rotates were based on the idea that this would be detectable, for example in the winds that it was thought such movement must create. At the equator, the speed of rotation is about 1000 miles per hour. If an aeroplane moves through the atmosphere at this supersonic speed, there is always a supersonic bang. So where is the bang that the Earth's movement should create? In fact, of course, there is no bang because the atmosphere rotates with the Earth. So, to a first approximation, there is no relative speed of the Earth with respect to its atmosphere. The problem seems even greater when we consider the speed with which the Earth is orbiting the sun, which is about 67,000 miles per hour. Again, the answer is that the atmosphere moves with us at this speed. In both cases, we don't notice the movement because there is no movement relative to what would make us feel the movement. Only when we compare the Earth's position relative to objects in the sky do we see the effect of the movement. These observations already take us well on the way to understanding Special Relativity. The key point is that we can only detect relative movement.

But consider this. It is nevertheless true that the Earth is moving through space. If space is the fixed structure that we learnt about in Euclidean geometry, then surely there must be a way of detecting whether or not we are moving through this structure. Since the atmosphere moves with us, we won't notice this by measuring the speed of sound. But what does light move through? Clearly it must be capable of moving through essentially empty space, otherwise we would not see the stars. Since we are moving with respect to space, we should be able to detect this movement by measuring how fast light travels when it moves with the Earth's movement compared to how fast it moves when travelling in the opposite direction. There should be a difference.

Another way to think of this is to imagine that space is filled with something through which objects travel. People called it the ether. If that was so, then there would be a privileged frame of reference in the universe: it would be the one in which movement through the ether is zero.

The experiment to test this idea was conducted by Albert Michelson and Edward Morley at Case Western Reserve University in the USA in 1887. The answer was a big surprise. There is no difference between the speed of light measured in any direction. The experiment has been repeated many times with ever-increasing accuracy and always with the same result. The conclusion is that there is no ether. There is therefore no privileged frame of reference.[21]

This is startling. Think about it and you will appreciate just how big a revolution this set in train. If space is not filled with anything that could form the basis of a frame of reference then the 'centre of the universe' is nowhere, or perhaps everywhere. Remember Cardinal Nicholas of Cusa, who as early as 1440 wrote: 'Thus the fabric of the world will have its centre everywhere and circumference nowhere.' It took more than 400 years for the world to catch up with his insight.

It is not certain whether this experiment was the trigger for Einstein's Special Theory, but the theory certainly provides an explanation for the result and so it became a major experimental proof of the theory.[22]

The fact that the speed of light is the same in all directions and that there is no privileged frame of reference leads to some counterintuitive consequences. When two objects are moving with respect to each other, let's call them A and B, distance and time in B must be perceived by A to be changed. Similarly, B will perceive A's distances and times to be changed. This is the phenomenon that leads to the famous space traveller example. A space traveller leaves Earth and travels a long distance at a very high speed relative to the Earth, and then returns to Earth. The space traveller will have aged in what will seem to him to be a normal way, but he will find that people on Earth have aged even more. By contrast, people on Earth will have experienced ageing at a normal rate and will think that the space traveller has discovered the secret of longevity! To them he would appear young.

These consequences lead to the fact that there will be no absolute measure of simultaneity. If time seems to change at different speeds depending on the relative velocities of objects, then we no longer have a privileged frame of reference to which to refer all events in their time sequences. An event that precedes another one at a distance away from it can appear to follow it if we change our observer position and relative speed. Think of three objects all moving with respect to each other: A, B and C. Suppose also that events occur in B and C that seem to be at the same time to A. Depending on how they are moving with respect to each other, they will not appear to be simultaneous to B or C. Their order in time will be different to different observers. Time and space are therefore no longer absolutes. They can contract and dilate according to the position and velocity of who is observing the events that occur.

It is important to note that these contractions and dilations of space and time do not allow the central rule of causality to be broken, even though their order in time can appear differently to different observers. The distance between the two events will always be such that, if the event on B is the cause of the event on C, no observer could consider C to be the cause of B.

There are many other surprising consequences of Special Relativity. One of the most important is that the speed of light becomes an absolute limit. However much we accelerate an object with mass, it will never achieve the speed of light. Only an object without mass can do that. Mass and energy also become inter-convertible according to Einstein's famous equation: $e = mc^2$.

I have summarised rapidly some of the main consequences of the Special Theory of Relativity. But this is not intended to be a textbook on the theory itself. Readers who wish to understand more deeply should read other texts – as suggested at the end of this book. The real purpose of this chapter is to prepare you for the rest of this book. Just as Einstein's relativity theories upset some common perceptions about space and time, mass and energy, light and gravity, so we will find that the consequences of extending the principle of relativity to biological processes also upsets many common ideas about causation and the relations between genes and organisms and their environments. If you find some of the consequences of the principle of relativity to be surprising or even shocking, you are not alone. Most people found the consequences of each stage of applying the principle to be surprising. The important point is the recommendation I made at the beginning of this chapter. Adopt the eyes of an inquisitive explorer. Don't hold on to your pre-conceptions unnecessarily.

Einstein's General Theory of Relativity

The Special Theory applies to the 'special' case of frames of reference moving at a constant speed with respect to each other. Each frame of reference can obey the rules of Euclidean geometry. In the special theory what we are abandoning is the idea that space is filled with something, the ether, that enables us to know whether we are at absolute rest or moving. But for each frame of reference, space can be treated successfully with the usual (which means Euclidean) rules of geometry. The angles of a triangle, for example, will always add up to 180°. A right-angled triangle will always obey the square rule for the lengths of its sides – the sum of the squares of the short sides equals the square of the long side.

In his General Theory, Einstein incorporated gravity. In doing so he also abandoned the assumption that space must be Euclidean. In this version of the theory, an object that is feeling the effects of gravity moves in a space-time continuum that is influenced by the masses on which the gravitational field depends.

General relativity has its most astonishing consequences at large scales when we are dealing with hugely massive bodies like clusters of galaxies. The way in which gravity is incorporated into the theory is best represented by saying it structures space-time around its field of influence. Space is therefore no longer represented as Euclidean. Space-time is like a sheet that can be bent and deformed according to the forces that act on it. This is just a two-dimensional way of representing what must be happening in four dimensions. Interestingly, mathematicians, such as Bernhard Riemann in 1854, and others even earlier in the nineteenth century, had already proposed non-Euclidean curved geometries as mathematical entities. It must have seemed that such geometries were only of mathematical interest.[23] But it happens often that new mathematics eventually finds applications in the real world.

This move in applying the principle of relativity is the most fundamental so far. Space-time is no longer independent of the objects within it. Perhaps even those objects can be represented as distortions of space-time so that they also are not independent of space-time. The dependence works both ways. At bottom (whatever we mean by that word – we will come back to that question later) objects may be just a set of formations in four-dimensional space-time that require even more metaphysical dimensions to represent what is happening. Some modern theories that seek to unify the great theories of physics, such as string theory, use precisely this kind of speculation. A very different approach is taken in a relatively little-known theory of the relativity of scales, which goes so far as to abandon the idea that space-time is continuous. Further reading on these fascinating questions at the very frontier of physics can be found in the endnotes for this chapter.

Can We 'Feel' the Consequences of Relativity?

If you react to Einstein's Special and General Theories of Relativity in the way I did when I first learnt about them, you may also follow the logic step by step, and assent to the conclusions with your head. But what about one's heart? Can people 'feel' its consequences? The early tests of the

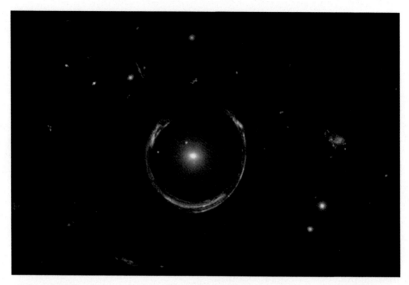

Figure 1.4 A remarkable example of gravitational lensing predicted by the theory of Relativity. The luminous red galaxy in the centre of this image, LRG 3-757, is almost exactly in front of a much more distant blue galaxy whose light is bent by the gravitational 'lens' to produce an almost perfect circle (source: Lensshoe_hubble.jpg: ESA/Hubble & NASA). For a colour version of this figure, please see the plate section.

theories were remote from everyday experience. Measuring the constancy of the speed of light was certainly not an everyday experience! One of the first tests of the bending of space-time by large objects was the observation during a solar eclipse that a star may appear near the edge of the sun that shouldn't have been where it seemed if space were Euclidean. There is just enough bending by the sun's gravity to make that observation possible. But that also was open only to the few who could make those observations accurately at the time of a total eclipse, and even they had to make several attempts to capture such a rare event.

But we now live in a privileged time. Thanks to the Hubble telescope up there in space and the ready availability of its images online, we can see for ourselves some of the wonderful discoveries that allow one to 'feel' the bending of space-time. A massive gravitational object can bend space-time and the light flowing through in a way comparable to a lens.[24] The most dramatic example of this effect is the distortion of light from a distant blue galaxy by a luminous red galaxy to produce the horseshoe image shown in Figure 1.4. In this case, there is almost perfect alignment of the distant blue galaxy, the red galaxy and the Earth to enable the distortion to produce an almost perfect circle. More frequently, the alignment is not that perfect and the result is a number of smeared arcs. A full horseshoe shape is itself rare. But had the alignment been even more perfect we would have seen a full circle (Figure 1.4).

There are even more wonderful discoveries that have come from the Hubble Space Telescope. The last section of this chapter concerns the Hubble deep field views of the distant universe and how they may be used to estimate the size of the visible universe. That estimate will prove very useful to our story in Chapter 3.

Hubble's Deep Field Views

This chapter began with the star constellation *Ursa Major*, and it ends with it. When astronomers were choosing a part of the sky to enable the Hubble to look at the very distant universe, they chose a very tiny region (only one-24-millionth of the whole sky) in order to point the Hubble at it to collect the light for a period of ten days. There are only four stars from our galaxy in this area, so to the naked eye and even most telescopes, the area seems empty, completely black. The Hubble was therefore looking all the way through our galaxy into far deep space. The area chosen is just above the 'plough' of *Ursa Major*, near where the bear's tail joins its back, which is the part of the formation that resembles a plough. The resulting image is amazing. Far from the sky being empty, it shows at least 3000 galaxies, many of them extremely faint and distant. The light from some of them has taken around 13 billion years to reach the Earth. In fact, the Earth did not exist when the light set out on its journey. The Earth is estimated to have been formed about 4.5 billion years ago (Figure 1.5).

Hubble's fabulous views take us to near the edge of the visible universe. They also enable a calculation to be made. If the universe looks roughly the same everywhere, which seems to be true, then by calculating the amount of matter from the total number of galaxies in the field of view and then multiplying it by 24 million we can estimate the total number of galaxies in the visible universe. That gives about 200 billion. Estimating the number of particles (e.g. protons, which form the hydrogen atom nucleus) in a galaxy at about 10^{69}, the total number of particles

Figure 1.5 The Hubble deep field view of one-24-millionth of the sky showing numerous galaxies right to the edge of the visible universe (source: apod.nasa.gov/apod/ap140605.html). For a colour version of this figure, please see the plate section.

in the universe turns out to be about 10^{80} particles.[25] That is a huge number, which would take most of this paragraph to write out fully, with 80 zeros. In Chapter 3 we will compare this number to the number of possible interactions in biological systems.

Conclusions

Relativity theory is inevitably associated with the name of Einstein – correctly so, although others, notably Poincaré, Lorentz, Gauss and Riemann, also developed many of the key mathematical ideas, and many others have done so since. Physicists, however, also refer to the general *principle* of relativity. By this they mean the distancing of our ideas from presumptions for which there is insufficient justification. So, many physicists also talk, completely correctly, about Galilean relativity or Newtonian relativity. Both Galileo and Newton were applying relativistic ideas, although they did not use the word relativity.

I think we should go even further. As shown in this chapter, each of the stages of trying to find the centre of the universe was an application of this very general principle of relativity. The stages are: first, abandon the idea of the Earth as flat while all the rest of the universe is spherical; second, abandon the idea that the Earth is the centre; third, abandon the idea that the sun is the centre; fourth, abandon the idea of any privileged frames for movement – movement is always relative; fifth, abandon the idea that there could be a centre – all frames of reference are equivalent; sixth, abandon the assumption that space-time is Euclidean.

I also imply that there are more applications of the principle to come in the future. Some physicists have already proposed extensions of the principle of relativity – for example, to the relativity of scales, a theory proposed by the French astrophysicist Laurent Nottale. The rest of this book explores applications of the principle of relativity in what I call Biological Relativity, which shares important ideas with the scale relativity theory of Nottale.

A second implication of this chapter is that there is also a philosophical principle of relativity. The discoveries of quantum mechanics forced us to distance ourselves from the idea that the universe is a vast piece of determinate clockwork. It has also forced some re-thinking concerning how we can know the world 'as it really is'. These implications will be explored more fully in the last chapter. But the lesson is worth bearing in mind throughout the book.[26]

Notes

1 Job 9:9.
2 http://en.wikipedia.org/wiki/Chinese_constellations
3 The seven stars of the big dipper are sometimes taken to be the *Septarishi* (seven sages) in Indian literature.
4 http://en.wikipedia.org/wiki/Principle_of_relativity

5 The first 'cosmological constant' was introduced by Einstein to enable his equations to produce a steady-state universe. If he had not done this he would have predicted an expanding universe. For a readable review of this problem and the cosmological parameters in general see Penrose, R. (2004) *The Road to Reality: A Complete Guide to the Laws of the Universe* (Jonathan Cape, London; pp. 772–778). Also, the wonderful little book by the Astronomer Royal: Rees, Martin (1999) *Just Six Numbers: The Deep Forces that Shape the Universe* (Weidenfeld and Nicolson, London).

6 Einstein, Albert (2010) *Relativity: The Special and the General Theory* (trans. Robert Lawson) (Ancient Wisdom Publications, Peoria, AZ).

7 In this case I have used the word 'uniquely' because 'privileged' would give the wrong impression. The idea was not that the Earth was a privileged 'centre' of the universe. On the contrary, the flat Earth was regarded as at the *bottom* of the universe, with an even worse place, Hell, below it.

8 See Cormack, L.B. (2015) Myth 2: that before Columbus, geographers and other educated people thought the earth was flat. In *Newton's Apple and Other Myths About Science*. R.L. Numbers and K. Kampourakis, editors (Harvard University Press, Cambridge, MA; pp. 16–22). The Wikipedia entry on this issue is also very clear: https://en.wikipedia.org/wiki/Myth_of_the_Flat_Earth

9 Rosen, Edward (trans) (2004) [1939] *Three Copernican Treatises: The Commentariolus of Copernicus; The Letter against Werner; The Narratio Prima of Rheticus* (second edition, revised) (Dover Publications, New York).

10 An epicycle is a circle whose centre moves round the circumference of another circle. If movement around this smaller circle is faster than the movement around the main circle then it would be possible to explain why some planets seem to travel backwards compared to their expected path.

11 Nicholas of Cusa, *De docta ignorantia*, 2.12, p. 103, cited in Koyré, A. (1957) *From the Closed World to the Infinite Universe* (Johns Hopkins Press, Baltimore, MD; p. 17).

12 *Dialogo sopra i due massimi sistemi del mondo* 1632.

13 Drake, Stillman (1978) *Galileo At Work* (University of Chicago Press, Chicago).

14 http://en.wikipedia.org/wiki/Laplace%27s_demon

15 Quoted and referenced in Fara, P. (2015) Myth 6: that the apple fell and Newton invented the law of gravity, thus removing God from the Cosmos. In *Newton's Apple and Other Myths About Science*. R.L. Numbers and K. Kampourakis, editors (Harvard University Press, Cambridge, MA; pp. 48–56).

16 Address to the British Association for the Advancement of Science.

17 Dirac, Paul (1930) *Lectures on Quantum Mechanics* (Princeton University Press, Princeton, NJ). Cox, Brian and Jeff Forshaw (2011) *The Quantum Universe: Everything That Can Happen Does Happen* (Allen Lane, London).

18 This is a highly technical field of research using several approaches to the question of how to harness quantum states effectively for computer memory and logical operations. See, for example, Harty, T.P., D.T.C. Allcock, C.J. Balance, *et al.* (2014) High-fidelity preparation, gates, memory, and readout of a trapped-ion quantum bit. *Physical Review Letters* 113; DOI: 10.1103/PhysRevLett.113.220501, which achieves the longest coherence time for a single physical qubit, together with highest precision manipulations of a single qubit. Also see Ballance, C.J., V.M. Schafer, J.P. Home, *et al.* (2015) Hybrid quantum logic and a test of Bell's inequality using two different atomic isotopes. *Nature* 528:384–386; DOI: 10.1038/nature16184, which achieves highest precision 'quantum logic gate' between two qubits, and a test of the quantum mechanical 'Bell's inequality' for different-species of atoms. These results are important steps on the way to the construction of practical quantum computers.

19 Saeedi, K., S. Simmons, J.Z. Salvail, *et al.* (2013) Room-temperature quantum bit storage exceeding 39 minutes using ionized donors in silicon-28. *Science* 342:830–833.

20 For a review of possible quantum mechanical effects in biological systems, see Melkikh, A.V. and A. Khrennikov (2015) Non-trivial quantum and quantum-like effects in biosystems: unsolved questions and paradoxes. *Progress in Biophysics and Molecular Biology* 119:137–161.

21 Over a century later, there is a fascinating twist to this story. A modern variation of the Michelson–Morley experiment has been developed in attempts to test an important prediction of Einstein's General Theory of Relativity. This is that light should be influenced by gravitational waves when objects of large mass distort space-time.

These experiments (called ADVANCED VIRGO in Europe, ADVANCED LIGO in the USA) use two tunnels at right angles and around 3–4 km in length to detect the exceedingly small effects of gravitational waves. To achieve this test, the apparatus needs to be trillions of times more sensitive than that used by Michelson and Morley. See www.ego-gw.it/public/virgo/virgo.aspx. On 11 February 2016, while this book was being finished, an announcement was made that gravitational waves had been detected for the first time. Abbott, B.P. *et al.* Observation of gravitational waves from a binary black hole merger. *Physical Review Letters* 116:061102.

22 The Michelson–Morley experiment was first interpreted by the Dutch physicist Hendrik Lorentz, who proposed that bodies change their magnitude when moving, which is the Lorentz transformation. Lorentz's equations were adopted by Einstein, who showed that the experiment and the equations were consistent with relativity theory, which gave the Lorentz transformation an 'incomparably more satisfactory' interpretation (Einstein, *Relativity: the Special and the General Theory*, p. 53). In his 1905 paper, Einstein does not reference Michelson and Morley. The implication is that the experiment was not itself used to formulate the Special Theory of Relativity, but was seen later by him to be neatly explained by it. For a historical analysis of the relation of the Michelson–Morley experiment to the Special Theory of Relativity, see Arabatsis, T. and K. Gavroglu (2015) Myth 18: that the Michelson–Morley experiment paved the way for the special theory of relativity. In *Newton's Apple and Other Myths About Science*. R.L. Numbers and K. Kampourakis, editors (Harvard University Press, Cambridge, MA; pp. 149–156).

23 A relatively simple way to grasp the idea is to imagine measuring the 'shortest distances' between points on the surface of a sphere. They will always be the distances along curved lines that form part of circles whose centre is the centre of the sphere. This is why, in two-dimensional projection maps of the Earth, aircraft seem to be travelling distances that are longer than the apparent shortest distance between the points.

24 Nottale, Laurent (1990) Gravitational redshifts and lensing by large scale structures. *Lecture Notes in Physics* 360:29–38.

25 This estimate is based on the visible (normal) matter. If we accept that this is only about 5% of the total energy and matter, including

dark energy and dark matter, then we could increase the estimate by one or two orders of magnitude, to 10^{81} or 10^{82}.

26 In writing this chapter, I would like to acknowledge discussions and debates with many scientific, mathematical and philosophical colleagues on the topics of relativity and quantum mechanics over many years. I would also like specifically to acknowledge a lovely book, *La relativité dans tous ses etats*, by the French astrophysicist and specialist on gravitational lensing, Laurent Nottale. Sadly, it exists only in French, but if your French is good enough you will benefit as I did from reading it. My ideas on the general principle of relativity owe much to his book. Nottale also has a book in English, *Scale Relativity and Fractal Space-time*, but this is not a translation of the smaller book. It is also highly mathematical and well beyond the scope of most lay readers. Nottale is the inventor of the theory of scale relativity, which has influenced my thinking about biological relativity since we share the same instinct to say that there is no privileged scale. Laurent Nottale and the French systems biologist Charles Auffray have explored the possible biological consequences of this idea (Auffray, C. and L. Nottale (2008) Scale relativity theory and integrative systems biology: 1. Founding principles and scale laws. *Progress in Biophysics and Molecular Biology* 97:79–114. Nottale, L. and C. Auffray (2008) Scale relativity theory and integrative systems biology: 2. Macroscopic quantum-type mechanics. *Progress in Biophysics and Molecular Biology* 97:115–157). It is important to note, however, that my use of the principle of relativity in this book does not depend on Nottale's specific theory of fractal space-time. The inspiration from his work lies in extending the principle of relativity to biological scales.

2

Biological Scales and Levels

There are only molecules – everything else is sociology.
Jim Watson (Nobel Laureate, author of *The Double Helix*)[1]

The Sense of Scale

In Chapter 1 we got a sense of the immensity of the known universe. Let's now go in the opposite direction, down towards the smallest scale in living organisms. We will see that it takes almost as many scale changes to go down to the lowest level as it did to go all the way up to the whole visible universe.

I was a student in the 1950s when the first electron microscopes were introduced in biological research. A normal light microscope can magnify up to about 2000 times.[2] By using electrons as the beam instead of photons we can increase this magnification to at least ten million. To enable the electrons to form a meaningful image they must pass through only very thin sections of material, so we cannot use electron microscopes for living cells. That is a serious limitation, but it is balanced by the fact that we can drill down to the molecular level. This is the way in which the British scientist Hugh Huxley saw for the first time the individual molecules called actin and myosin. These are long protein filaments, and he showed that they must slide over each other when a muscle contracts. He was even able to see the small molecular protrusions called cross-bridges that enable this sliding movement to occur. Another Huxley, Andrew Huxley (not related), was able to make the same

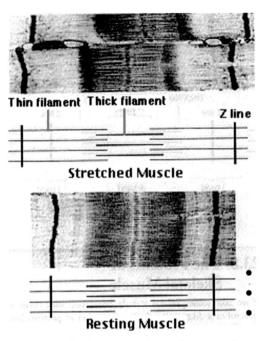

Thin filament Thick filament

Z line

Stretched Muscle

Resting Muscle

Figure 2.1 The first electron micrograph images of the arrangement of protein filaments in skeletal muscle, together with diagrams showing how the filaments slide along each other when the muscle changes length (Huxley and Hansen, 1954).[3]

deduction from experiments using clever light microscopy. These discoveries led to the famous sliding filament model of muscle contraction (Figure 2.1).

As medical students at University College London, where Hugh Huxley was working, we had the opportunity to see where he had his electron microscope. It felt like entering a holy of holies. We were allowed to enter one by one through a sliding door into the dark room where the precious instrument was housed to see for ourselves the beautiful arrays of the filaments. It has always seemed to me surprising that this work was not honoured with a Nobel Prize. Andrew Huxley did receive one, but for his work with Alan Hodgkin on nerves.

At the magnification required to see the muscle filaments the cell that housed them would have covered the whole of the square mile or so around the university. That dizzying fact brought home to me the vast ranges of scales on which biological organisms exist. This chapter explores that range and what exists at each scale.

Scales and Levels

Before we extend the principle of relativity to biological systems, we need to know what living systems are made from and how they are organised. Is the material of living systems special, or is it just the same as the rest of the material in the universe? If it is just the same, then what defines the difference between living and non-living? We can begin to answer those questions by determining what the components are and how they are organised to produce the activities of living systems.

There are various ways in which the components of an organism can be classified. One of these uses the concept of levels of organisation. These levels can be arranged in an ascending order of size: atoms, molecules, networks, organelles, cells, tissues, organs, whole-body systems and the whole organism. Each level can be viewed as assembling the components of the level below, and then itself forming a component for the level above. We can also go beyond the organism to add populations, species, clades and the physical environment, which can also be classified into various levels. To some extent this order is also an order of scale. Cells, for example, work at a scale much larger than that of molecules.

Cells themselves are so small that we need a microscope to see them. Yet, they are enormous compared to atoms and molecules. Each cell contains trillions of atoms. The same relative difference of scale applies to cells compared to a whole human body. There are many trillions of cells in a whole human body. The scales therefore are very different. One way to imagine the difference of scale between the level of atoms and that of the whole body is to note that there are more atoms in the body than stars in the visible universe. When you contemplate the deep field view of the Hubble (see Chapter 1) and the sheer immensity of the known universe, remember that from the viewpoint of a single atom, our bodies are also as immense as the universe. The differences of scale are truly vast and,

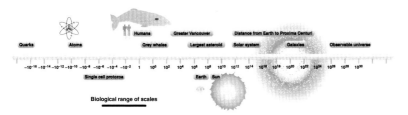

Figure 2.2 Scales from the tiniest subatomic particles to the whole observable universe. On the logarithmic scale used here, man lies roughly midway between the smallest subatomic particles and a whole galaxy (http://hendrix2.uoregon.edu/~imamura/123/lecture-1/lecture-1 .html). For the colour version of this figure, please see the plate section.

as it happens, organisms are roughly in the middle with about as many scales below them as above them (Figure 2.2).

Levels and scales are not the same. Scale refers to the dimensions and boundaries of a chosen subset of nature. Level refers to (often roughly) distinct forms of organisation. The difference can be appreciated by noting that some organisms lack some of the levels I have listed. For example, in unicellular organisms the cell and the organism are the same. Organisms exist at a large range of scales, from the smallest bacterium[4] at around 0.4 μm to a huge elephant of about 4 m. This range is ten million. Clearly, then, the scale of a bacterium is very different from that of an elephant, even though both are complete organisms: 10^{21} bacteria could fit into the body size of an elephant. This may surprise you. Even more surprising, there is more biomass on the Earth in the form of bacteria and other unicellular organisms than in the whole of the rest of the animal and plant kingdoms. There are more bacteria in your and my bodies than we have human cells. They are essential to our functioning as humans. This is a form of inter-dependence of organisms, more usually in co-operation than in competition (Figure 2.3).

Scale is a more neutral description than level since it does not depend on organisation, even though different forms of organisation occur at the different scales. By contrast, the concept of level depends precisely on what we identify as forms of organisation. The level of cells depends on the form of organisation we call a cell, which is viewed as being above the level of molecules and, in multicellular organisms, below that of the whole organism.

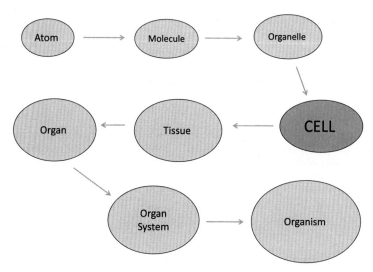

Figure 2.3 Levels of organisation in biology. The level of the cell is highlighted as particularly significant. The reasons for that view will be explored in Chapter 4 (Adapted from http://easysn.wikispaces.com/ANATOMY).

The concept of level is therefore a metaphysical one using the metaphors of up and down. In fact, the 'lowest' levels – atoms and molecules – are everywhere in the body. Just as there is no centre of the universe, there is no 'real' central or privileged level of the body, and no literal bottom or top either. If you see an analogy here with Cardinal Nicholas of Cusa's fifteenth-century insight (Chapter 1) 'the world will have its centre everywhere and circumference nowhere', you have already begun to understand a major message of this book.

By the way, this point is quite different from the question of whether there are *anatomical* centres, tops and bottoms. Of course, there are, at least by convention. Gravity ensures that my head is conventionally regarded as above my feet. This would not be a natural metaphor in deep space. Yet it would still be correct to say that my body is at a larger scale than my cells or my molecules.

One of the problems of the way in which twentieth-century biology was frequently presented to the public was to mistake the metaphors of up and down – and many other metaphors – for reality. This has also been a problem for scientists themselves, since many were taken in by their own rhetoric, and many still are. The result has been a biological science version of the illusion that there is a centre of the universe. In biology this illusory centre is often supposed to be DNA molecules, as the repository of genes. When discovering the structure of DNA, Francis Crick famously announced to his drinking companions in a Cambridge tavern that he had discovered the 'secret of life'. The director of his Institute, Max Perutz, was rather more careful than Crick when he said that DNA was the 'score of life'. That is more correct since a musical score does nothing until it is played. DNA does nothing until activated to do so.

The illusion has become so strong that many people think we know exactly what a gene is and how genes control the body. Those are also illusions. Recent experimental work in biological science has deconstructed the idea of a gene, and an important message of this book is that it has thereby dethroned the gene as a uniquely privileged level of causation.[5] As we will see, genes, defined as DNA sequences, are indeed essential, but not in the way in which they are often portrayed. They are passive, not active, causes. That important difference will become clearer in later chapters of this book.

Metaphors are colourful and convenient, and it would be hard for biological scientists to work without the idea of levels and the associated metaphors of up and down, but we must keep in mind that these are just metaphors and that nature does not need to follow the way in which we classify it. What is convenient to enable us to think about and classify nature is not necessarily a sure guide to the way in which nature works. The application of the concept of levels to biology is just as necessary but also just as metaphysical as the application of Euclidean geometry to space. It can be just as incomplete or even wrong. With those points in mind, let's look at the nature of the various biological levels.

Atoms and Ions

Organisms are constructed from some of the same atoms as physical systems. Atoms consist of a nucleus made from the particles protons and neutrons. Protons carry a positive charge, while neutrons carry no charge. Different atoms contain different numbers of protons and

Elements in the periodic table occurring in living systems

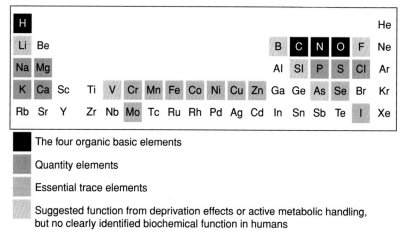

The four organic basic elements

Quantity elements

Essential trace elements

Suggested function from deprivation effects or active metabolic handling, but no clearly identified biochemical function in humans

Figure 2.4 Part of the periodic table showing the elements found in living organisms in the lower-weight region of the table (adapted from: https://en.wikipedia.org/wiki/Template: Periodic_table_(nutritional_elements)). For a colour version of this figure, please see the plate section.

neutrons, collectively called nucleons. That is what makes the difference between the simplest atom hydrogen, with one proton, and, for example, sodium, with usually 23 nucleons, 11 of which are protons. It is the proton number that determines the classification since it also determines the number of electrons, which are needed to match the charge of the protons, and the electrons in turn determine the chemical properties. The proton count is also called the atomic number.

Atoms have chemical properties that depend on the atomic number. The Russian scientist Dmitri Mendeleev, working in St. Petersburg, showed in 1869 that atoms could be arranged into a table displaying periods and groups such that their properties could be predicted from their position in the table. This system is called the periodic table and it remains the one used today to classify atoms, although we now know about twice as many types of atoms as were known in Mendeleev's time. The principle of his table remains valid. The atomic number and its position in the table give a good guide to the kinds of interactions an atom can take part in (Figure 2.4).

The number of nuclear particles of stable atoms can range up to nearly 238 (92 protons), which is the number at which we reach the unstable atom uranium.[6] The most common atoms found in living systems are much smaller. Just six relatively small atoms – hydrogen (1), carbon (6), nitrogen (7), oxygen (8), phosphorus (15) and sulphur (16) – account for 97% of the weight of the human body. These atoms are used over and over again in almost all the molecules of living systems, including DNA, RNA and proteins. Hydrogen provides the largest fraction of atoms in an organism, around 63% in a human. It is also the most common element in the universe.

In much smaller quantities there are the atoms that form the charged ions: sodium (11), potassium (19), chlorine (17), calcium (20) and magnesium (12). These five atoms account for nearly all the remaining 3% of matter in an organism. Other larger atoms are present only in very small (trace) quantities. One of these is iron (26), which is essential in proteins like haemoglobin, which transports oxygen around the body. A few others, like manganese (25), cobalt (27), copper (29), zinc (30), selenium (34), molybdenum (42) and iodine (53), are also essential to form some of the enzymes (active protein catalysts) of the body, and they also occur only in trace quantities. But they may have been essential for the origins of life on earth (see *RNA and Other Early Worlds?* in Chapter 4). Other elements are found in even tinier amounts, but are not thought to play a physiological role. Many of them are harmful.

The body is therefore formed of very large numbers of a few of the most common small atoms in the universe, with a tiny sprinkling of a few larger atoms that play a role in enabling certain key molecules to perform their special functions. Think of a pancake made from flour, water, eggs and oil, which form the great bulk of the dish, and then a small seasoning of pepper and salt and other spices. You then have a culinary comparison. A little spice goes a long way. A few enzymes using heavy metal atoms are sufficient for living systems based on DNA to function.

The positively charged nucleus of an atom is balanced electrically by a cloud of surrounding smaller particles called electrons. Remember from Chapter 1 that these particles, like all particles, also function as waves and that the energy jumps that electrons can make are quantal. A cloud of electrons is therefore a good description: we cannot know where any particular one is until it is detected by an instrument. All the chemical interactions depend on the outermost electrons in the cloud, the outer shell. Electrons in this shell can jump out of or into it. For example, common salt – sodium chloride – forms by the sodium atom losing an electron to become a positive sodium ion, while the chlorine atom gains one to become a negative chloride ion. Sodium chloride forms most of the salt in the sea, and since organisms first evolved in seawater the cells of the body exist in saltwater, reflecting the amounts that were in the sea during the origin of life. The salt composition of the inside of cells, however, is very different, as we will see later. The main positive ion inside is potassium. Sodium is held at a much lower concentration. Calcium is even lower still. These details are very important for the functioning of living cells. The gradients between the inside and outside can be used as sources of stored energy.

As I have already noted, most of the universe is made of hydrogen, either as the atom itself or just its nucleus – a single proton forming the hydrogen ion. The hydrogen ion also plays an important role in organisms. The amount of free protons in any region of the body determines what is called its acidity. Acidity (also referred to as pH) is an important property of living systems and has to be controlled within narrow limits for everything to function well. So, too, is temperature, which represents the amount of kinetic (moving) energy the atoms have. High temperature represents a lot of movement, low temperature less movement. In many organisms, temperature is also carefully controlled within narrow limits. As we will see, these controls are part of what make living organisms special. It is not the atoms themselves which make organisms special, it is how they are arranged and controlled that defines a living organism. The quantities of many other ions and molecules are also controlled within certain limits. Organisms can detect what the levels are and use this to determine whether to raise or lower the levels.

Physicists have revealed that the particles of atomic nuclei, i.e. protons and neutrons, are themselves composed of even more fundamental particles. Biologists are generally not concerned with these since we are fairly certain that they never operate independently of atoms in biology. Nevertheless, this fact is important when considering the question of whether there really is a 'lowest' level from which all biological causation begins. We need to remember that below neutrons and protons there are quarks, gluons, etc. This is an inconvenient fact for those who think that it is somehow obvious that the bottom (causal?) level in biology is that of atoms or molecules. Why should we stop at those levels? And who knows whether physics has already reached what may be the 'real' bottom level.

Nor is it evident how any such 'reality' would help. As we saw in Chapter 1, at the micro level physics is described in quantum mechanical terms where the concept of reality itself is in question. The full reasons for saying that there is no 'centre' and no 'bottom' will be dealt with in Chapter 6. Those reasons for the change in viewpoint that I am arguing for are best appreciated by first knowing some of the important experimental observations on the basic components.

Before we leave the level of atoms, we should note that the particles that can fuse to form atoms do so in the fusion reactions that can occur in stars, which did not appear to light up the sky until thousands of years after the origin of the universe (the 'big bang'). A universe that did not form stars would not have formed living systems as we know them. We are, literally, 'stardust'. So is the Earth we inhabit. On current views it and we form part of a vast process of unfolding from a tiny and highly dense origin, which would have been completely invisible to the naked eye. But then, no eyes could have existed at that time. Eyes, like the rest of our bodies, are made from molecules, which are made from atoms. At the time of the big bang no atoms existed.

Molecules

Atoms are combined in molecules. Examples of these include water, which forms more than half of the mass of a human. Others include the atmospheric gases we breathe: oxygen, carbon dioxide and nitrogen. There are also many atoms that are in the form of ions, which are atoms that have lost or gained one or more of their surrounding electrons. These include sodium, potassium and calcium ions, which are positively charged because they have lost one or two electrons. They also include hydrogen ions, which are therefore protons, since hydrogen has only one electron to lose, and the nucleus is a single proton. Then there are negatively charged ions, like chloride. Sodium chloride is common salt. But in the water environment of the body it divides into the positively charged sodium and the negatively charged chloride. I don't need to produce a complete list of all the molecules – what you will understand from this is that most of the body is made of the same kind of 'stuff' that we find outside the body.

But organisms also include many molecules that are not usually found outside them. Some of these are molecules that many readers will know about from popular science books and media programmes. These include proteins, DNA and RNA. You will learn much more about RNA in this book and you may have to relearn some of what you have learnt about DNA. Proteins, and the DNA and RNA that form templates for the manufacture of proteins, are molecules called polymers that are constructed from many small molecules arranged in long chains. Amino acids form the chains in proteins. Individual nucleic acids called nucleotides form the polymer chains in DNA and RNA.[7] Nucleotides also play a major role in energy usage in organisms, where ATP is a kind of universal energy currency, and in signalling, where cAMP is used.[8]

There are also special oily (fatty) molecules called lipids. They play an essential role in the formation of cells, which are the fundamental functional units of an organism. The earliest organisms were single cells. In fact, most organisms alive today are also single cells. Lipids are essential for forming the oily membranes at the cell surface and at the surface of many objects inside the cell. The oily nature is essential to the way in which lipids function. Water and oils tend to separate, as any good cook knows when a delicious blended sauce breaks up and has to be re-blended. The same process occurs in living systems. The oily lipids stick together to form extremely thin films surrounding the watery interiors of cells and organelles. They naturally form thin membranes, just as oil naturally spreads in a thin film on water. In biological membranes, there are two thin films forming a double layer. That enables the membrane to separate two water-based spaces since each layer of the double membrane has a water-loving surface facing out to the water phase, with an oily surface connecting to the oily surface of the other layer to form the very thin interior of the membrane.

The fatty molecules do not therefore form an oily drop, as usually happens when we mix oil and water. It is hard to see how an oily drop could form the basis of living systems since such a drop would keep the chemicals that dissolve in oil separate from those that dissolve in water. What is needed is a separate water space where the special reactions of living

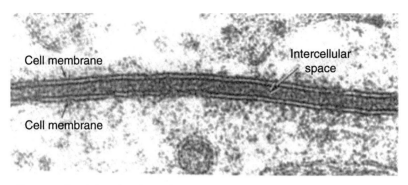

Figure 2.5 Double layer (lipid bilayers) forming the membranes of two adjacent cells with an intercellular space between them. Each bilayer is about 5 nm thick. This is far smaller than the wavelength of light (400–700 nm), which is why the bilayers cannot be seen under the light microscope. The development of electron microscopes made their visualisation possible (source: Bloom & Fawcett 1994)[9].

systems can take place and where oily molecules can sit in membranes and allow their active sites to be accessible to other molecules from the water phases (Figures 2.5 and 2.6).

An important discovery on lipids was made in the 1960s by the Cambridge scientist Alex Bangham. He showed that lipid bilayer vesicles (small spheres a bit like a tiny cell – see Figure 2.6) form spontaneously when mixing dried lipid with water. The lipid bilayers are fluid: molecules like proteins sitting in the bilayer can move around. The bilayers are also self-healing. If broken, the layer will reform naturally. These properties will be important when we look in more detail at cells in Chapter 4.

Then there are many relatively small molecules called metabolites, hormones and transmitters. The metabolites move around and interact to form the biochemical networks that generate energy, and to perform many other functions essential to life and reproduction. The hormones and transmitters perform the functions requiring communication between different parts of the organism. We will meet all of these later. Most of them also dissolve in water

These special biological molecules must have evolved in the enclosed space since they do not naturally occur outside living organisms. Many of them can be made artificially, and they can also be isolated from the body and investigated scientifically in the same way as other molecules. The rules that describe how they can be formed and how they interact are the same as those of chemistry in general. We no longer think that there

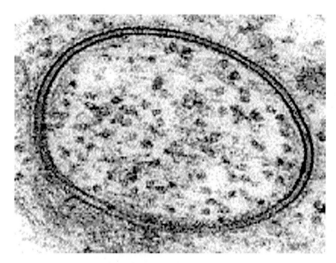

Figure 2.6 Transmission electron microscope image of a lipid vesicle. The two dark bands around the edge are the two leaflets of the bilayer. As in Figure 2.5, the bilayer thickness is around 5 nm. Historically, similar images confirmed that the cell membrane is a bilayer (Wikipedia, Sandraamurray, File:Annular_Gap_Junction_Vesicle.jpg).

is a special kind of matter or energy in living organisms. The nineteenth-century question of whether there exists a special 'vital energy' only found in organisms was largely settled in that century. The answer is no. However, we will see later in this book that a distant relation of the concept of 'vital energy' has made a comeback. Properties of biological systems include behaviour that only the system can display. As I will show, these systems properties play a role rather similar to the original concept of vital energy.

Molecules as Systems

In this book I will frequently refer to systems. A system is a combination of components that interact with each other. It is characteristic of systems that their properties can be hard to deduce from knowledge of the components. Often, those properties are surprising. This is not just true of organisms. It is also true for molecules.

As an example, consider the most abundant molecule in an organism, water. It consists of two common atom types, two hydrogen atoms and one oxygen atom. At the temperatures of living organisms, those are gases. Hydrogen needs to be cooled to -253 °C to become liquid. Oxygen also needs to be cooled to a very low temperature, -183 °C, to form a liquid. But as everyone knows, water is a liquid at a much higher range of temperatures between 0 °C and 100 °C, and can remain liquid at even lower temperatures when other molecules are dissolved in it. Some of those molecules are very adept at lowering the freezing point. This is the way in which fish and other sea creatures can survive temperatures below 0 °C and even survive freezing. They can do so because some dissolved molecules can also stop frozen water from forming the kind of sharp crystals that damage cells.

The special properties of water do not end with the temperatures at which it is liquid. They also include special ways in which it forms a crystal as it freezes. In solid water – that is, ice – the molecules are further apart and so ice is lighter than water. That is why ice first forms on the surface of a lake rather than at the bottom. It is also why icebergs float in the sea. For organisms that require liquid water to survive that is very convenient, as is the fact that water is heaviest at about 4 °C.

A deep enough frozen lake or pond will have water at about 4 °C at its bottom, where organisms can continue to flourish. Ice and water are also good insulators, so the bottom of the pond will be protected to a large extent from very low atmospheric temperatures. This is the reason why large lakes outside the polar regions rarely freeze completely. Since the ice is at the surface, it will also melt fastest when the atmosphere becomes warmer.

Water is also a good solvent, which means it mixes well with many other molecules, other than the oily ones. It is ideal therefore as a base substance within which other substances can interact. It is hard to imagine a living system without the equivalent of such a solvent since the biochemical interactions that form many of the networks of the body use water as the medium within which the movements and interactions can take place.

Other molecules carry some water molecules around with them since their charged nature attracts water just as one magnet attracts another. There is a cloud of water molecules around an ion, for example, which is said to be hydrated. In fact, any charged molecule will attract a water shell and be hydrated.

These are just some of the special properties of water. Even though we know that water is composed of hydrogen and oxygen, the properties of the molecule itself are clearly characteristic of the combination. They are not the properties of the individual components in isolation. Nor are they what we might imagine if the properties were a simple mixture of those of its components. The components interact and this interaction is essential to the properties of the whole.

There is a common misunderstanding about this kind of interaction, where the interaction itself changes the properties of the individual components. A theoretical chemist might object to this idea by saying that once we know enough about the physics and chemistry of hydrogen and oxygen we can use equations to predict their properties in water. That is doubtless true, but it is still correct to say that those properties emerge from the interaction. A key feature of Biological Relativity developed later in this book is that components are always and necessarily constrained by the whole. That does not mean that the components do not continue to obey the equations of quantum mechanics and of interatomic forces. This is a good example of what I mean by a systems property, and it is a phenomenon we will encounter frequently in this book.

It is clear that, even at the 'bottom' level of the molecules in organisms, we are dealing with systems. Molecules are also systems. Remember this point when I come later on to discuss DNA. DNA is also a system! It is not just a sequence of nucleic acids. As a system it interacts with other systems in the organism. Also, it is controlled by the rest of the organism in ways that will be described in Chapters 4, 7 and 8.

The properties of water are so special, and so important to the development of life, that we naturally encounter another big 'why?' question of the kind raised at the beginning of Chapter 1, when we contemplated the night sky and the 'why?' question about the universe. Water looks as though it was designed with life in mind. That is a discovery we have made using scientific investigation. Can the same methods be used to answer this 'why?' question, and all the other 'why?' questions we will encounter in this book?

Many books on science ignore this kind of question by treating it as an inappropriate, even meaningless, non-scientific question. Those that do address it often claim that science can answer it, and that the answer is very simple: there was no design, and no designer. I will show that it all depends on what we mean by an answer to such questions. The principle of relativity can also be applied to what we can know. We will see that goals can be properties that emerge during the process of evolution.

Goal-directedness is therefore also a relative property. Goals are always relative to some particular context, internal or external to the organism. Goal-directedness is traditionally called teleology, but many scientists have avoided using the term since Francis Bacon's great work *Novum Organum* in 1620 outlining the inductive method, which formed the basis of the reductive approach in science. That has been interpreted to mean that scientists should avoid teleology, famously referred to by the nineteenth-century German physiologist Ernst Wilhem von Brucke as 'the lady without whom no biologist can live. Yet he is ashamed to show himself with her in public.' Monod and Jacob even adopted the term 'teleonomy' to refer to 'apparent' teleology.[10] But interpreted as the simple and verifiable statement that the behaviour of a system is goal-directed, there is nothing 'apparent' in the behaviour so I see no reason to avoid the word teleology. The objection raised by many scientists is that teleology must entail belief in an overall purpose in nature and therefore belief in an ultimate intelligence, or god. In a relativistic context, however, this is not the case. Teleology as purpose in the context of a particular organism–environment interaction makes perfectly good scientific sense; it is verifiable. An engineer has no difficulty in performing tests to determine whether a man-made machine exhibits purposive behaviour towards a goal. Biologists can perform similar tests to show that a living organism has goals. The whole universe – or whatever might lie 'beyond' it – may have no knowledge of my goals as a living being. They are nonetheless real, as are the goals of a rabbit, an amoeba or a bacterium.

The reason why we can identify goal-directed behaviour in living organisms is that we can understand the processes by which the behaviour emerges. Those processes can be investigated and analysed scientifically. We will see examples of that kind of understanding in subsequent chapters. Some 'why' questions can therefore be most certainly answered by scientific investigation.

Could we do the same for the 'why' question in relation to a molecule like water? The answer at the moment is no. We simply do not understand the processes by which the fundamental properties of matter emerge that enable the characteristics of water that are advantageous to life.

This issue will be taken up again in subsequent chapters and in Chapter 9 in the context of the relativity of epistemology.

Networks

In this section, by networks I mean biochemical networks: the ways in which the molecules of an organism interact with each other within their liquid environment. Later in the book, I will use the term network a little more widely.

Molecular networks can be just two or three interacting molecules or they can be thousands, millions, billions or trillions. There is no limit to the size of a network. Ultimately, the whole organism is a vast molecular network. The problem is that, at that scale, we have little hope of understanding it fully as a molecular network. Networks become complicated with even five or ten elements and they certainly become difficult to calculate and understand when they involve, say, 100 elements.

This is an inescapable mathematical fact, but also a practical one. Imagine a knot made by intertwining two lengths of string. Even with just two strings we can make knots that are difficult to unravel. Add some more strings and very rapidly we arrive at knotted networks that defy even the cleverest at unravelling knots. Understanding networks is a little like the problem of understanding knots.[11]

What do we do when faced with such a problem in everyday practical life? One way to deal with it is to throw the knot away and start again with fresh string. Metaphorically, we 'cut the Gordian knot'. Is there an equivalent approach in understanding the 'knotted' complexity of organisms? The answer is yes, and it all depends on levels.

One of the roles of the concept of levels in biology is that when too many elements are involved at one level, it helps to try to understand the system at a higher level. Just as the detailed workings of the processors in a computer may not be necessary to understand how the computer implements a particular program, so it is not necessary to understand everything at a molecular level in order to understand a biological function. In fact, I would go further. What may be impossible, because of sheer numbers of elements and their interactions at one level, often becomes clear at a higher level. We should try to answer questions at the level to which they are most appropriate and then use that insight to probe down and up towards the other levels. This is the approach that the Nobel laureate Sydney Brenner characterised as 'middle-out'.[12]

Many biological functions are brought together at the level of the cell. The rhythm of the heart, for example, is best understood at that level. So are many other rhythms, such as circadian (daily) rhythm. They depend on networks that involve cell properties as well as properties of the individual molecules.

That is one reason why the concept of levels is important in biology. Biology is not alone in this regard. Physicists use the same approach when they refer to temperature, which is a global property of a system of elements, rather than to the individual movements of all the elements. Pressure, pH (acidity) and electrical potential are similarly global properties of ensembles.

Organelles

The very simplest organisms, called prokaryotes and protists, have fairly minimal structures inside their cells. More complex organisms have another level in addition. The interior of their cells is packed with what look like cells within cells. We call these organelles. They include the nucleus, where the DNA normally resides to form the genome. The nuclear DNA is also closely associated with nuclear proteins (called chromatin) to form chromosomes. The nucleus contains many other molecules. It is not just a bag of DNA. Possessing a nucleus is the defining characteristic of what are classified as eukaryotes, the kinds of cells found in your body and in those of most multicellular organisms. The only cells in our bodies that do not normally contain a nucleus are the red cells of the blood. They are cells that have lost their nuclei, do not need to provide energy for it and so are more efficient in their oxygen-carrying role. They survive for around 100 days and, of course, they can never divide and have progeny.[13]

The part of the cell outside the nucleus is called the cytoplasm. It also contains organelles, sometimes many of them. In animals, there are mitochondria, which are the energy factories of cells. Muscle cells which use a lot of energy are packed full of them. The equivalent organelles in plants are called plastids. There are also protein factories, the ribosomes. Some of these are surrounded by membranes just as a cell is. Don't bother to learn all these names just yet. This section is to introduce the fact that cells in organisms like you and me, and in most other multicellular organisms, contain other components that have some of the characteristics of cells. You might well ask whether, if there are cells within cells, perhaps these originated themselves from formerly independent cells. That is exactly what we now think in some cases at least. This fact has major implications for evolutionary biology, which will be explored in Chapters 4, 7 and 8.

Cells

Galileo revolutionised astronomy with his telescope and his observations of the moons of Jupiter. Antony van Leeuwenhoek from Delft in Holland did the same for biology with his microscopes and his observations on the fine structure of cells and organisms. His work was published from 1673 in the then young journal of the Royal Society, *Philosophical Transactions*, and it revealed a world that until then was completely unknown. Together with the work of Robert Hooke, who first coined the word 'cell' for the basic structure of living organisms, he discovered the cellular structure of plants, animals and microorganisms. Leeuwenhoek was also the first to identify an organelle inside cells, the vacuole. Vacuoles are temporary organelles used for getting rid of excess water. They grow progressively and then discharge their water and any other components by fusing with the surface cell membrane so that a passage can open up to the exterior. Lipid spheres in cells can fuse and divide, just like soap bubbles. In a sense they are bubbles, but in water rather than in air.

Leeuwenhoek's discovery of microorganisms was initially received sceptically by the Fellows of the Royal Society, largely because it required a major revision of ideas on the origin of life. The theory popular at the time was that the tiniest forms of life, what we now call microorganisms, could form spontaneously in the right conditions. His work established that organisms could be single cells and that they divide to produce more cells. Life comes from life, except of course for the first living systems (Figure 2.7).

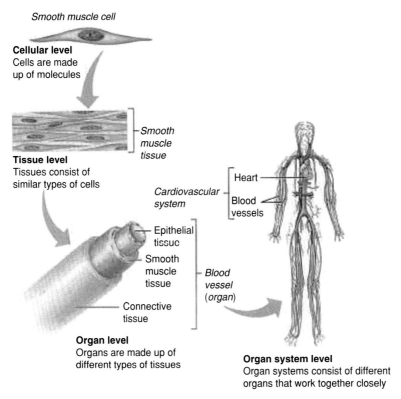

Figure 2.7 Images of cells, tissues, organs and organ system levels using smooth muscle as an example (*J Physiol.* 2014 Jun 1; 592: 2375–2379). For a colour version of this figure, please see the plate section.

That was an extremely important discovery, but it leaves open the question of the origins of life on Earth. We no longer think that life continues to originate from non-living material. The origin was an event, or more probably a long series of events, around four billion years ago. It does not happen today, probably because, once established, living systems changed the Earth in ways that may prevent it happening again. Living systems have completely changed our atmosphere, for example, which did not have sufficient amounts of oxygen molecules when life was first formed. The atmosphere today would be toxic for early forms of life, and the early atmosphere would be toxic for us and many other forms of life today. Organisms are continually changing their environment. The idea that a particular environment forms a niche, within which certain organisms can thrive, is only part of the picture. Those niches were in turn often created by previous generations of organisms. There is a circular form of causality between organisms and their environment. Circular causality is a central feature of the theory of Biological Relativity. It also applies to the interaction between organisms and their DNA, and between DNA and the environment. In all cases the interaction is both ways. Organisms are not isolated systems.

Hooke's reason for using the word cell was that, under the microscope, the cells in a plant resemble the rows of cells for monks living in a monastery. The analogy is a good one. A cell encloses a space with an outer layer, the wall. In the case of a living cell, this is formed by an oily membrane composed of lipids. Like a room, the cell has doors and windows. These are the proteins in the membrane that we call channels or transporters, which allow substances to cross between the inside and the outside of the cell.

I said earlier that what characterises living organisms is not the atoms of which they are made, but how these are controlled. The cell membrane with its proteins is one of the key controllers because the channels are selective. They only allow certain substances to cross. The substances kept inside can therefore be very different in their quantities from outside. The difference can be very large. For example, the amount of calcium ions inside a cell can be very small indeed, as little as 1/10,000th of the concentration outside. The mechanisms by which these large differences are established and controlled give living systems independence, a kind of freedom from all the vagaries of the outside world.

Moreover, in addition to creating an internal environment that is different from the outside, it is possible in a small space to produce and control changes in the quantities of substances. To use the example of calcium again, this ion has been developed in living systems to be the controller of cell movement. By increasing the calcium level by around ten times (to about 1000 times smaller than outside) the cell can move from rest to moving. And it can relax back again quickly when the calcium level is reduced. This is the process that enables us to move our muscles rapidly and under fine control. That would be impossible without cells. A large local variation in an ion or molecule is best achieved within a confined space. Even within the cell itself, local changes in confined spaces can be much larger than the change in the whole cell. The intricate structure of cells forms the basis of very fine control mechanisms.

Control in living organisms depends on the formation of compartments. Structure is one of the keys to understanding control in living systems, which is why their anatomy – the study of structure – is important. The information contained in 3D cell structure is comparable to that contained in the genome.[14] The nature and origin of cells will be addressed in Chapter 4.

Tissues

For the great majority of evolutionary history, no tissues or higher levels existed. The cell was the complete organism. Moreover, most of the organisms alive today are single cells. In fact, as I noted earlier, most of the cells in your body and mine are single-cell organisms: the bacteria that make our gastro-intestinal systems work so well match or even outnumber the cells in the rest of our bodies. Co-operation of this kind between species is widespread. Nature is not always, nor even usually, 'red in tooth and claw'.

There are many forms of co-operation between cells. Some species, like the amoeba *Dictyostelium*, have life cycles in which they are sometimes free-moving single cells, while at other times the cells aggregate to form a slime mould in which the cells are no longer free to move independently. This can be seen as a primitive version of a multicellular organism. Multicellular organisms are cells that have given up independent existence. There are also 'organisms' that are two species co-habiting to their mutual benefit. Corals and their co-operation with algae are a good example. Amoeba-like organisms called foraminifera (which means organisms with holes) also co-operate with algae that live in the holes in remarkable examples of what we call symbiosis: two species co-operating. The bacteria in our gut also form just such an example. We, too, are symbionts. As I have already noted, all eukaryotic cells owe their origin to symbiogenesis.

Aggregation produces remarkable changes in the properties of the individual cells. The whole ensemble determines how the individuals behave. This can be seen in the slime mould, which behaves like a fungus or plant with a fruiting body that enables dispersion of its seeds, called spores, to occur. The spores eventually become new single-cell organisms and the cycle can then begin again. As with aggregation of atoms to

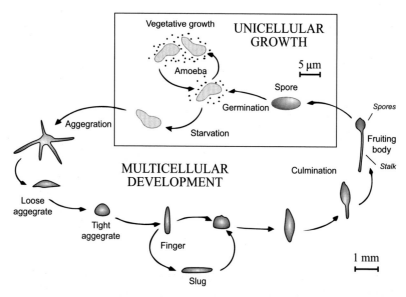

Figure 2.8 Life cycle of amoeba *Dictyostelium discoideum* (Tijmen Stam, drawn by user:Hideshi, then converted to SVG by IIVQ; from: https://commons.wikimedia.org/wiki/File:Dicty_Life_ Cycle_H01.svg (reuse permitted under GNU Free Documentation Licence and CC-BY-SA Licence)).

form molecules with utterly different properties from the components, this is an example of a system, with properties that are characteristic of the system (Figure 2.8).

In multicellular organisms a tissue is a group of cells that function together to form the building blocks of organs and systems. Cells of a particular type aggregate to form sheets (hence the name tissue) and blocks of various shapes, just as slime moulds form as aggregates of the individual amoebae. Once again, the interaction between the cells changes their properties. Moreover, the way in which this happens can vary enormously even in the same organism.

When we classify the types of cells in our own bodies, we can distinguish around 200 different types according to the properties of the tissue in which the cells find themselves. They can be as different as bone cells, using calcium to form hard bone, and heart cells, using calcium to control the heartbeat. Even the single atom calcium is made to behave differently according to the system within which it finds itself. In one it is a building block of a solid architectural feature of the body; in the other it is a lithe controller, jumping around between cell organelles at high speed to control the movement of the heart. In skeletal muscles it does so to enable, for example, a musician to play accurately and rapidly. Think of a guitarist playing a tremolo piece or a pianist executing fast arpeggios: on all of those finger muscles millions of calcium ions are moving around very rapidly in small fractions of a second. From a single-atom viewpoint it is impossible to explain this. Calcium is constrained by the cells in which it finds itself to behave completely differently. The ways in which wholes constrain parts is a key feature of the concept of Biological Relativity.

All of the 200 cell types show differences that are as striking as this. Liver cells perform biochemical functions that nerve cells do not, while nerve cells have their own specialisations. Nerve cells communicate tiny electric signals around the body, so helping to control its functions rapidly. If you see a squid darting backwards at great speed to avoid a predator while also ejecting a cloud of black ink to confuse the predator, the speed with which that can occur is a consequence of a remarkably large nerve cell that conducts the electric impulse to activate the huge mantle muscle. This is jet propulsion, and it is an avoidance control system that would make any manufacturer of a fighter aircraft proud. One moment the squid is in front of a hungry predator, and the next moment as the predator opens its mouth to attack all it encounters is a cloud of ink with no idea where the tasty squid has disappeared to. In the case of some cephalopods, the jet mechanism can even power long flights completely out of the water and through the air. When nature *is* 'red in tooth and claw' it pays to be that quick, and adding a 'disappearing act' to the trick can be the difference between life and death.[15]

There are many more examples of such specialisations in cells. But now we come to the truly remarkable fact about these specialisations. All of these cells in our body have almost exactly the same DNA in their genomes! That should already warn you that it cannot be the DNA alone that determines what happens in cells, tissues and organs. The same genome can be interpreted in completely different ways. Some organisms even incorporate this principle into their life cycles. A caterpillar and the butterfly it transforms itself into have exactly the same genome. Some have speculated that this may also be an example of symbiosis. The caterpillar and the butterfly are so different that, conceivably during evolution, they started off as different species.[16] Even DNA can co-operate!

Organs

The word 'organ' makes us think automatically of our own bodily organs: hearts, lungs, livers, kidneys, etc. But to a musician it would mean the instrument with many pipes found in churches and concert halls. Actually, both uses of the word share the Latin and Greek origin, which means an instrument, something used for a particular purpose. The concept of a functional goal is an essential part of the concept of an organ.

An organism is a complete system of organs, each used by the organism to do a specific set of jobs. What provides that purpose is often seen as a contentious question. Biologists and philosophers have had great difficulty answering it. How can purposive, living organisms develop from inanimate purposeless atoms and molecules?

As we develop the idea of Biological Relativity it will become clear that part of the problem is that the question of purpose or function depends on the level we are considering. It has no meaning below the level at which the relevant function is integrated. As an example, consider the heart in vertebrate bodies like our own. In the context of the organism as a whole it is now obviously true that one purpose of the heart is to pump blood around the body to keep all the cells and tissues supplied with oxygen and energy. That has been clear ever since William Harvey's experiments in the seventeenth century demonstrating the circulation of the blood and the role of the movements of the heart in ensuring that circulation. But it is equally clear that at the level of the atoms and molecules that form the heart, there seems to be no such purpose. Those atoms and molecules are the same as others in the body that do not form the heart. The purpose is not therefore intrinsic to the molecules. If one insists that all causation in biological systems is molecular, then there is no 'real' purpose anywhere in the body. This is the origin of the claim by some scientists that nature is blind to purpose and that evolution must also be so. Yet, we also know that the heart would not exist if it did not serve its purpose.

Resolving this apparent conflict between molecular and higher-level views requires a fully relativistic theory of biology in which questions can be answered at a level that permits the question itself to be meaningful. There is no point in trying to answer a question at a level where it is meaningless. Different questions are appropriate at the different levels.

In particular, it isn't possible to understand the interaction of multiple components if you only ever study a single one in isolation.

In the hierarchy of levels we are discussing here it is natural to place organs at a high level, above cells and tissues. But before we explore that aspect, it is important to recognise that all organisms have organs, even if they are single-cell organisms. That is why the various components inside cells, like mitochondria, ribosomes and vacuoles, are called organelles, which means little organs. The nucleus is also an organ in this sense. Even more importantly for our story, so is DNA. Like other organs, it is used by the organism for a particular purpose. It was the Nobel Prize-winning plant biologist Barbara McClintock who first described the genome as an 'organ of the cell' in 1983. The full reasons for her insight will become clear in Chapters 6 and 7 of this book.

The word organ therefore includes the concept of purpose, what the organ is for.

There are organs, in this general sense, at all levels of biological organisation (and notice the appearance of 'organ' also in that word). This should warn us that nature doesn't need to pay too much attention to our ways of classifying her. When we come to develop the full principle of Biological Relativity, this fact will become very important.

Whole-Body Systems

The term 'systems biology' rapidly became a buzz phrase during the first decade of this century. Before the year 2000 it was rarely used, although the word 'systems' was used in a related sense in 'systems physiology', and in 'systems engineering', 'systems theory' and other related disciplines. Why did this development happen so quickly? In part, it was a reaction to a great achievement, the sequencing of the complete human genome. There were a lot of surprises arising from that work. One of them was that people realised that sequencing the genome would not be sufficient to answer the question 'what is life?'. A new approach was needed. Having smashed Humpty-Dumpty into billions of pieces – all those molecules forming genes and proteins – the time had come to work out how to put him together again. This necessarily requires building up rather than breaking down, integration to complement reduction. This is what systems biology is about. But the systems approach is not new. The word 'systems' has a long history in physiology. The various organs of the body form parts of whole-body physiological systems within which their function is defined and implemented. The new generation of 'systems biologists' has tended to ignore this use of the term 'system', and to focus on the biochemical networks inside cells. It is an important message of this book that this is a limitation. Many of the properties of biochemical networks cannot be fully understood without understanding their functions within the large systems extending throughout the body. Those systems contain the networks, cells and organs to achieve the goal of the system. In this part of the chapter I will briefly summarise the characteristics of these whole-body systems and how individual organs form parts of those systems.

Circulatory System

This is also an example of extensive co-operation. To remain alive, all cells need to be within about 50 μm of flowing blood from which they can receive oxygen and to which they give up carbon dioxide. As tissues grow in size so must the vascular system. A process called angiogenesis ensures this. The vascular network literally grows into all the new tissue. Individual cells in the system must be receiving control signals from the tissues and organs of the body.

The heart is a central part of the circulatory system. It pumps blood into arteries, which then feed the blood into smaller and smaller branches as they fan out across the whole body. The smallest branches are called arterioles (a diminutive term, rather like the use of the word organelles for the tiny 'organs' inside cells). These lead into the tiniest vessels of all, the capillaries. These are so small that the cells of the blood, such as red cells carrying the oxygen to the tissues, only just squeeze through. The Italian anatomist Marcello Malpighi first observed capillaries in the lungs of a frog in 1661. Like Antony von Leeuwenhoek, he was one of the first to use microscopes to investigate the tiniest structures in living organisms. He published his book, *De polypo cordis*, in 1666 in which he describes the composition of the blood. Four decades previously, in 1628, William Harvey had published his great work, *De motu cordis*, in which he demonstrated the circulation of the blood, but he had to infer the existence of capillaries. He never saw them. Yet his experiments and calculations showed that they must exist. Malpighi's work confirmed their existence and that they were too small to be seen by the naked eye.

After flowing through the fine and extensive network of capillaries, the blood is collected through a set of vessels called veins that show a structure opposite to that of arteries. Instead of dividing and subdividing again and again, they unite again and again, just as streams unite to form rivers. Finally, the largest veins carry the blood back to the heart to be pumped through the body again.[17]

Each part of this system serves its function within the system as a whole. Moreover, the properties of the system as a whole determine many aspects of its components. The size and power of the heart as a pump, for example, matches what is required in order to generate the pressure needed to overcome the resistance of the branching network of vessels. This is the context in which we can say that the heart, the vessels and the blood serve their purposes.

Respiratory System

In a similar way, the lungs are organs that serve their function within the respiratory system. They are the first step in the intake of oxygen by the body and the last step for the expulsion of carbon dioxide. They are served by a fine network of capillaries where the gas exchange with the blood occurs. In fishes and other aquatic creatures, the same function is served by the gills. In organisms like us, a separate blood circulation, also pumped by the heart, services the lungs. These gas exchange organs form a network including the chemical processes within the cells of the body that use the oxygen to create energy and produce the carbon dioxide to be expelled from the body. The complete system is called the respiratory system, and it also forms the context in which it makes sense to talk about the functions of each component.

Lung cells and tissue grow into the space they occupy. When doing this, both the lung and the vascular system show an interesting systems property that can ensure the most efficient space-filling structure.[18] In mathematics such structures emerge from simple recursive formulae that can generate highly complex and beautiful structures like the Mandelbrot set.[19] Similar algorithms must be capable of describing how lung and vascular tissues radiate and branch as they grow.

Urinary System

In the simplest single-cell organisms, water is evacuated from the body by temporary organelles called vacuoles. Clearly this cannot be suffi-cient in multicellular organisms where water and other substances have to be expelled through a separate system. The central organ in this sys-tem is the kidney, which filters the blood in a highly refined structure of tubes that control what is retained and what is excreted. Transport and exchange through the walls of the tubes are the processes by which this control is achieved. The output of this organ is then sent to the equiv-alent of the vacuole, the bladder, to accumulate before being expelled. This also is a whole-body system since it is controlled by small molecules called hormones that are produced elsewhere in the body. These controls constrain the cells and molecular structures of the urinary system to co-operate in the elimination of waste products.

Digestive System

Similar processes of transport and exchange occur in the digestive sys-tem. Unicellular organisms simply ingest their food by literally wrapping themselves around it and forming a temporary organelle, a kind of vac-uole, which carries the food such as a bacterium or algae into the cell to the organelles, called lysosomes, that contain the chemicals that enable the food to be broken down. These chemicals are proteins that speed up the reactions involved. They are called enzymes. In multicellular organisms the process is more complicated. Early in the development of the organism, a tube is formed running the whole length of the body, one end of which, the mouth, takes food in while the other end, the anus, passes out the waste. In between the entrance and the exit various kinds of specialised organs develop to break the food down and transport it across the wall of the tube to pass into the bloodstream. In humans, these specialised organs include the stomach and the intestines. Surprisingly, perhaps, the process of digestion is not achieved by the stomach and intestines alone. A wide variety of bacteria live in the digestive system and help the digestive process.

In addition to the digestive tube (oesophagus, stomach, intestines, colon) there are organs that perform specific functions connected with digestion. These include the salivary glands, the liver, gall bladder and the pancreas.

Nervous System

All organisms are sensitive to what is happening in their environment. Unicellular organisms have proteins in their surface membrane that can detect touch and movement. Mechanical sensitivity may well have been the first form of sensitivity as life evolved. An organism needs to know when it has encountered an object or another organism. Variations in chemical levels can also be detected. That is how an organism can detect where its food supply may be. The surface membrane of a unicellular organism is therefore its nervous system. The effects are transmitted immediately to the interior of the cell and can be coupled to networks of interactions that generate movement. This is a complete functioning system. There is a receptor on the cell surface, a transmitter to the interior of the organism, and an effector, which in this case is the mechanism that enables the organism to move. This three-component system is the basis of all nervous systems. There can be many kinds of receptors, many internal biochemical networks and many mechanisms of movement. Even bacteria have this kind of 'nervous system'. This is what enables them and all organisms to show goal-directed behaviour.

In multicellular organisms the link between the three components cannot be so direct. The relevant muscle cells will be located at a distance from the cells that sense the environment. The answer to this problem was the development of nerve cells. Nerve cells are like all other cells in most respects. They have their organelles: nuclei, mitochondria, ribosomes, etc. But they also have long extensions of themselves that reach out to interact with other nerve cells and with the effector organs, such as muscles and glands. Some of these extensions are like trees with many branches. The tree is called a dendrite, and it connects with the extensions of many other nerve cells to form networks of incredible density and complexity. The human brain contains tens of billions of nerve cells, each of which may connect to thousands of other cells. The number of possible circuits is effectively unlimited.

The problem of connecting with distant parts of the organism is solved by some of the extensions being very long indeed. We call these long extensions axons. They can be as long as the organism itself. There is a famous axon in the neck of a giraffe that travels all the way along the neck and back again. There are axons in your body that travel all the way from the top of your brain down through the spinal cord to connect to a nerve that sends a shorter axon to the muscles. Depending on the level in the spinal cord where the synapse occurs, this can be a distance of around 1 m. For a giraffe those nerve axons can be as long as several metres. With a diameter of around 10 μm, they can be 100,000 times as long as they are wide.

The advantage of nerve axons is that they can transmit very quickly. I referred earlier in this chapter to the nerve axon in the squid and other cephalopods that triggers the jet propulsion away from predators. The rapidity with which the transmission occurs is due to the unusually large size of the axon, which can be as much as 0.5 mm in thickness, and so visible to the naked eye. Most nerve axons in our and other vertebrate bodies, however, are much smaller and they achieve their rapidity of transmission through multiple layers of insulating lipid membranes. This is yet another use that evolution has produced for membranes. Membranes alone are very good insulators. Without channels almost nothing crosses them. Even water and respiratory gases pass through protein channels in membranes, not through the lipid layer itself. Without channels, therefore, circular layers of lipid membrane are so good as insulators that they can force electric current to flow along nerve axons and thus ensure rapid transmission over long distances.

Musculo-Skeletal System

Organisms have many ways in which movement can occur within them and to move themselves within their environment. Even plants display movement. Roots need to burrow to find water and nutrients, stems and trunks transport water and nutrients up and down from the leaves, while plants also twist and turn to achieve maximum exposure to light. Unicellular organisms show a bewildering array of motors of various kinds. Some swim by means of paddles called cilia attached to their surfaces. This may also be an example of evolution by symbiogenesis. Cilia may have originated from ciliate bacteria. Other kinds of cells move by pushing out a protuberance called a pseudopodium (literally a 'false foot'), using proteins within the cytoplasm to achieve the movement. If the pseudopodium finds an interesting object that could be food, for example, the whole organism can then move to ingest it by a kind of reverse process to that of forming a vacuole. Instead of using a lipid bag to expel fluid, it is used to ingest food.

At some point during evolution, some very useful long thread-like proteins developed called actin and myosin. These form the mechanical working parts of skeletal, cardiac and smooth muscles in our bodies. Actin and myosin are formed as long filaments with regular protuberances that are used to form bridges between them. Movement occurs by these bridges performing a kind of rowing movement so that actin and myosin can slide along each other, as we saw at the beginning of this chapter.

In vertebrates, muscles exert their actions through their connections to the bones of the body. The skeletal system serves at least two purposes. First, to give support to the whole body, which would collapse if it was composed only of soft tissues. Second, to provide the system of smoothly articulating joints that allow the muscles to move bones, much as a mechanical crane moves its parts. Nature invented pulleys, levers, articulations, lubrication and gearing long before the industrial revolution. It also knew how to multi-task. The interior of bones contains a remarkable organ: bone marrow is where blood cells and the cells of the immune system are made. As noted earlier, it is also a remarkable fact that calcium plays such a different role in bone and muscle. In bone it forms part of the molecules that make the dense crystal structure to form the solid part of bones. In muscles, as the calcium ion, it moves rapidly between cellular structures to control movement.

Immune and Endocrine Systems

The whole-body systems we have considered so far consist of discrete tissues and organs, although their effects are distributed throughout the body. The two systems we are now going to consider are inherently distributed. These are the immune and endocrine systems.

The immune system is the organism's defence department. When attacked by foreign bacteria or viruses or other substances, a part of the foreign object, called an antigen (antibody generator), triggers the production of proteins called antibodies that can bind to and neutralise the antigen. Since antigens can exist in many different molecular shapes and sizes, this requires a virtually infinite number of possible antibodies, far beyond what could be generated by using inherited DNA alone. How the cells of the immune system achieve this is fascinating and will be discussed in Chapter 7.

The antigen–antibody interaction is a version of a lock-and-key relationship. The structure of the antibody is a kind of negative template for the structure of the antigen, so that they click together chemically. Substances that do this naturally within the body itself are hormones and receptors. Receptors are proteins in cells whose structure enables them to click together with small chemicals – peptides – and proteins, called hormones, that act as signalling molecules to activate or inhibit various processes in the cells of the body. The hormones are produced by special cells in the body called endocrine glands. They form an important part of the multi-scale interactions between different levels of the organisation of an organism. Unlike the immune system, the endocrine system produces a finite number of hormones with fixed interactions with receptor proteins. By acting as messengers between different organs and systems of the body, they are essential to the integrative action of the body as a whole.

Integuments: The Body's Armour and Sensing System

Most of the simplest unicellular organisms are separated from the environment only by their cell membrane, although there also exist unicellular organisms that have specialised coverings, such as the foraminifera, which produce a solid shell usually made of calcium carbonate. The rocks from which the pyramids of ancient Egypt are built consist almost entirely of the compacted fossils of foraminifera.

Of course, many multicellular organisms, such as the crustaceans, also form hard shells. Other multicellular organisms form a skin of specialised cells. The skin in organisms like us is an extremely elaborate system, with many layers and components. In addition to forming a protective covering, the skin contains various kinds of sensory cells providing the sensations of touch, heat, cold and pain. The receptor parts of the specialised sense organs, such as eyes and ears, develop from regions of embryonic skin, while smell and taste receptors remain located in the inner surfaces of the mouth and nose. The protective function of the skin is extensive since it is involved in resisting pathogens, controlling heat and water loss and retention, and in healing wounds. In fact, the skin is continuously healing itself since new cells form to replace those that are sloughed off every day.

Classification of Systems of the Body

The classification of the organ systems of the body is not very precise. There is considerable overlap between the different systems. All of them use the circulatory system. The nervous system is distributed everywhere. Even the intestines have a 'brain' formed by a considerable number of nerve cells. Your gut instinct may well be less metaphorical than you might think! The heart, too, has a nervous network associated with it. The old idea that the heart was the seat of the soul may also not be quite so absurd as we have come to think.[20] The diffuse systems, immune and endocrine, are also everywhere in the body.

Does it make sense, therefore, to speak about particular organs, such as the heart and the liver? Of course, this is convenient from an anatomical viewpoint, but much less so from a functional viewpoint. The difference is nicely illustrated by the problems encountered in translating and interpreting between different medical traditions. Western medical science has been strongly reductionist in its approach. The word 'heart', for example means a particular anatomical structure with a well-defined boundary, essentially what we dissect out when removing a heart for transplantation. The oriental medical tradition uses the equivalent word, which is the Chinese character 心, to refer also to the system of which the heart is a part. This is a more integrative view.

The Organism as a Whole

All these systems function in the context of the organism as a whole, which forms the 'internal environment' of the cells, tissues and organs. The organism itself is a system. In the context of this book it also needs to be emphasised that it is an open system. Causal relations between scales do not stop therefore at the boundary of the organism. Those causal relations are also two-way: organisms create their environment to some extent. We as humans depend on that process for our very existence. We would not be able to survive in the atmosphere of the Earth during the first billion or so years of life here. We and all other oxygen-breathing organisms had to wait for other organisms to create the atmospheric oxygen that we depend on.

Beyond Organisms

Organisms are open systems. They continually exchange energy and matter with the environment, which is what we mean by an open system. Open systems cannot be understood without studying these exchanges with the environment, which itself is therefore a causal agency in the behaviour of the system.

More complex organisms interact with the environment in other ways too – for example, through social interactions within and beyond their own species. They create the environment of the future by creating new niches which in turn enables new forms of life to evolve to fill those niches. This creates chains of organisms of different sizes and interlocking forms of behaviour. The food chain is an obvious example. We would not exist without it. If humans ever manage to colonise another planet, they would probably have to transport with them the means of creating and maintaining a food chain.

Conclusions

Understanding the organisation of organisms in terms of scales and levels is an essential prelude to developing the theory of Biological Relativity. Twentieth-century biology tended to fragment into different disciplines: molecular biology, cell biology, physiology and so on up to ecology, the study of organisms in their context, and sociology, where the context is social. But this hides the fact that the really important processes that define living systems are interactions *between* the scales and levels. As we have seen in this chapter, although molecules have systems properties that would be hard to predict from the properties of their component atoms, and even harder to predict from the properties of fundamental particles, they are not living. All molecules in biological systems on their own are dead, including DNA. Even viruses, which are essentially DNA with some protective molecular structure, are hardly living outside the context of living cells. Within the context of life as we know it today, the minimal kit required to be a living organism is found at the level of a whole cell. This is important because in all organisms this is the level at which inheritance occurs. One of the important messages of this chapter is that a cell is vastly more than its DNA, and an organism is vastly more than a collection of cells. All of that 'vastly more' is passed on to subsequent generations. The consequences of this fact for biology generally and for evolutionary biology in particular are profound and we will explore those consequences in subsequent chapters.

The Nobel laureate Jim Watson was more right than he knew when he quipped: 'There are only molecules – everything else is sociology.' Even a molecule has a kind of sociology in its context, and organisms certainly do.[21]

The second conclusion is that the range of scales involved is simply enormous. At the beginning of the chapter I noted that to a single protein filament of a muscle cell, the cell itself would appear to be the size of a small city. An even more dizzying expression of the scales involved is to consider the smallest independent unit in cells, which is the proton (the centre of a hydrogen ion). Its diameter is about 1 fm (10^{-15} m). If we could magnify the proton up to the size of a cell, the edge of the cell would be beyond the edge of the solar system. I emphasise the vastness of these ranges because we can otherwise be misled by schematic diagrams that show the components of cells. To represent single molecules, such as individual proteins, we have to draw them *much* larger than they are for the diagram of the cell to be possible.

We are now ready to look at biological networks and how they work to enable the functions of living organisms to occur.

Notes

1 A similar view was expressed by Francis Crick in his book *Of Molecules and Men*. Essentially this is the view that life can be reduced to physics and chemistry. This was also the central issue between Max Perutz and Karl Popper following his 1986 Medawar lecture at the Royal Society (see Chapter 8). Popper tried to convince Perutz that he was wrong in thinking that a complete reduction was possible. The solution to this problem is to realise that molecules are

constrained when they are involved in a living organism, in ways that will be explained in Chapter 6. That constraint is a property of the system, not of the individual molecules. Popper was therefore correct.

2 This limit is set by the wavelength of light. There are some specialised applications in which the resolution can be increased to some degree, but never as much as by electron microscopy; see https://en.wikipedia.org/wiki/Optical_microscope#Surpassing_the_resolution_limit.

3 Huxley, A.F. and R. Niedegerke (1954) Structural changes in muscle during contraction: interference microscopy of living muscle fibres. *Nature* 173:971–973.

4 Viruses can be even smaller, but they are not independent organisms. They depend on the host organism to be able to reproduce.

5 Noble, D. (2011) Neo-Darwinism, the modern synthesis and selfish genes: are they of use in physiology? *Journal of Physiology* 589:1007–1015; Noble, D. (2015) Evolution beyond Neo-Darwinism: a new conceptual framework. *Journal of Experimental Biology* 218:7–13; Newman, S.A. (2013) Evolution is not mainly a matter of genes. In *Genetic Explanations: Sense and Nonsense*. S. Krimsky and J. Gruber, editors (Harvard University Press, Cambridge, MA; pp. 26–33, 288–290).

6 Uranium isotopes are not stable over very long periods of time, between 69 years and 4.5 billion years.

7 DNA and RNA are both nucleic acids consisting of strings of nucleotides. The difference is that the sugar part in RNA is ribose; in DNA it is deoxyribose.

8 ATP and AMP are formed from the nucleic acid adenine. ATP is adenosine triphosphate, AMP is adenosine monophosphate.

9 Bloom, W. and D. Fawcett (1994) *A Textbook of Histology*, 12th edition (Chapman and Hall, New York), figure 1.2.

10 The term teleonomy was first introduced in 1958 by Colin Pittendrigh to contrast with 'teleology', since the latter requires intention and foresight on the part of an agent responsible for the purpose. 'Teleonomy' was meant to apply when no such agent exists, the purposive behaviour was to be regarded as a natural consequence of physical processes without intentionality. My view is that

organisms with intentions have evolved and can therefore display teleological properties that can be assessed and measured, just as an engineer can do in the case of a purpose-built machine that exhibits goal-directed behaviour.

11 The analogy with knots is a good one. There is a branch of mathematics concerned with knot theory. See https://en.wikipedia .org/wiki/Knot_theory.

12 In the Novartis Foundation Symposium on *The Limits of Reductionism in Biology*, 1998.

13 Although red cells do not divide, there are cells that can do so without nuclei. Lorch, I.J. (1952) Enucleation of sea-urchin blastomeres with or without removal of asters. *Quarterly Journal of Microscopical Science* 93:475–486. Of course, cells that achieve this do not pass any DNA on to their progeny but they do pass on all the extra-nuclear cellular inheritance.

14 Noble, D. (2010) Differential and integral views of genetics in computational systems biology. *Interface Focus* 1:7–15.

15 Notice also that the squid's struggle for existence, its 'selfish goal', is based on extensive co-operation between its own parts. The metaphors of 'selfish' and 'co-operative' are level-dependent concepts. This fact is particularly relevant to genotype–phenotype relationships. Far from the selfishness of an individual being determined by its 'selfish' genes, it actually depends rather on its co-operative genes. This will become clearer when the language of 'selfish-gene theory' is analysed in Chapter 5.

16 Williamson, Donald *The Origins of Larvae* (Kluwer Academic Publishers, Norwell, MA). This theory is highly controversial; see http://en.wikipedia.org/wiki/Donald_I._Williamson.

17 The precise organisation of the circulation varies between species. In amphibians, reptiles, birds and mammals there is a separate circulation to the lungs, so the heart has two sides, one pumping to the lungs and the other to the rest of the body. In fish and in many other organisms the arrangement is different.

18 Technically, these are called fractal structures. The pattern of branching is the same at all scales, which means both that the smaller scales resemble the large-scale structure, and that space is filled very efficiently.

19 For examples of Mandelbrot sets, see http://en.wikipedia.org/wiki/
   Mandelbrot_set.
20 This statement requires a context to be properly understood.
   Physiologists have unambiguously established that the heart can
   function without attachment to the nervous system. The rhythm is
   generated within the heart muscle itself, as described in the next
   chapter. The nervous system regulates the frequency and the
   strength of the heartbeat through release of the transmitters
   adrenaline (accelerator and force augmenter) and acetylcholine
   (decelerator). The ability of the heart to generate its own rhythm is of
   course important in heart transplantation, when it is disconnected
   from its donor nerve network. My statement concerning 'seat of the
   soul' is intended not to question the well-established physiology of
   the heart and its intrinsic functions, but rather to warn against views
   of conscious experience that locate it entirely within the brain. See
   chapter 9 of Noble, D. (2006) *The Music of Life* (Oxford University
   Press, Oxford). Also see Noble, D., R. Noble and J. Schwaber (2014)
   What is it to be conscious? In *The Claustrum: Structural, Functional
   and Clinical Neuroscience*, J.R. Smythies, L.R. Edelstein and V.S.
   Ramachandran, editors (Academic Press, New York; pp. 353–363).
   From the viewpoint expressed in those chapters, conscious
   experience is a property of the body as a whole, one of the most
   important processes in a systems view of life.
21 See note 1.

3

Biological Networks

*In nature, there is no 'above' or 'below', and there are no hierarchies.
There are only networks nesting within other networks.*
Fritjof Capra and Pier Luigi Luisi 2014,
The Systems View of Life: A Unifying Vision

Networks are not Diagrams!

Years ago it was usual for many biological laboratories to display posters of what is called the metabolic network, showing the interaction connections between the different molecules in a cell. This was almost a badge of honour. Often lavishly produced by companies, they were initially quite useful. As we will see in this chapter, metabolic pathways are immense. But when our knowledge of them was restricted to a few key sub-networks it was possible to 'know' such pathways, just as a medical student is expected to know the anatomy of the body. As biochemical science rapidly grew in the later part of the twentieth century, that kind of rote learning became impossible. Nobody today could draw the complete known metabolic pathway poster from memory. They have become so complicated that even on large posters one needs a magnifying glass to view the relevant detail. The more 'complete' of these posters would now cover most of a large wall (Figure 3.1).

Could there ever be an end to this increasing complexity? The answer is no.

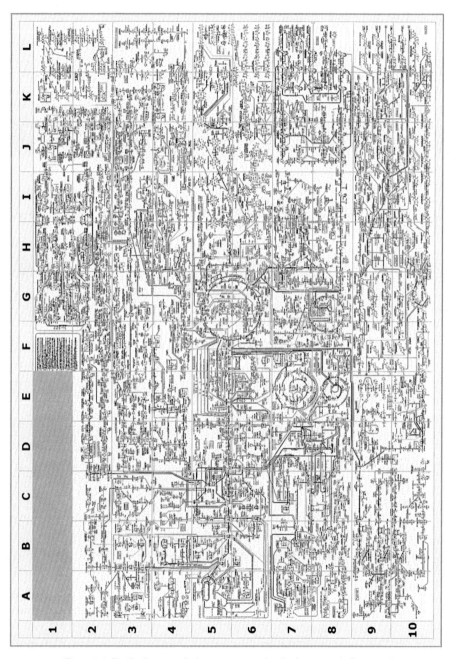

Figure 3.1 The Roche network chart as an example of a dense metabolic network poster (rotated).

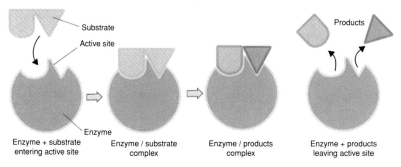

Enzyme + substrate
entering active site

Enzyme / substrate
complex

Enzyme / products
complex

Enzyme + products
leaving active site

Figure 3.2 Lock-and-key representation of the way in which enzymes work. In this case the two parts of a molecule to be broken up fit into the template formed by the enzyme, which then influences the molecule to make it easier to break the chemical bonds. The reverse process would describe how an enzyme brings two separate molecules together to make them more likely to join. This is how enzymes enable long strings of nucleic or amino acids to link together to form DNA or proteins (https://commons.wikimedia.org/wiki/File:Two_Substrates.svg). For a colour version of this figure, please see the plate section.

Even the most complicated of these diagrams are simply two-dimensional static maps of what is in fact a dynamic four-dimensional process. Furthermore, they are not multi-scale. They simply summarise our knowledge at a molecular level. As we will see, functional biological networks necessarily include interactions with higher levels. So, how should we look at networks?

Any or all of the components of living systems described in Chapter 2 can form networks. A network is simply a system of interactions. It can be as small as two components interacting with each other, for example to form a third component, but the word is usually meant to indicate many more. Biochemical (molecular) networks were introduced in Chapter 2. A good example is a metabolic pathway. Metabolic pathways consist of hundreds of different chemicals called metabolites. The interactions between them are speeded up by proteins called enzymes. It is thought that before proteins evolved this process was carried out by RNA molecules, which can also speed up reactions, though in less specific and much less effective ways. In both cases the process works by using the shape of the RNA or protein to bring the two metabolites close together, so reducing the energy barrier to interaction (Figure 3.2).

Each metabolite fits into the relevant shape of the catalyst. Proteins can do this more efficiently than RNA since there are many more

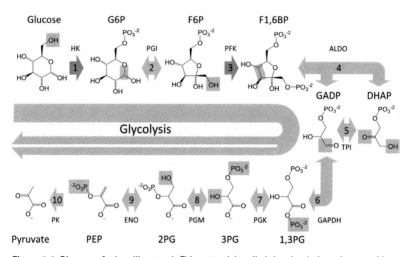

Figure 3.3 Diagram of a 'small' network. This network is called the glycolytic pathway and has ten links, each one of which is enabled by a particular protein enzyme. The links represent the processes by which glucose is broken down to produce energy. There is no need to remember the details of this diagram. The important details can be found on Wikipedia under the entries 'glycolysis' and 'enzyme'. What is relevant to this chapter is that even small processes in biological networks are complex in their details (from Wikimedia Commons, Thomas Shafee). For a colour version of this figure, please see the plate section.

specific shapes that can be made from 20 amino acids compared to just four nucleotides (Figure 3.3).[1]

Functional networks of this kind are the workhorses of biology. They mediate energy transformations, for example from sunlight and food, and so enable work, such as muscle contraction, glandular secretion or neural integration, to be done. Sunlight is the main source of energy in plants, where photons are absorbed in organelles called plastids. Food is the main source in animals, directly from other animals or ultimately from plants. The 'small' network in Figure 3.3 represents the steps by which the sugar glucose is used to create energy in the form of ATP, so it is one of the most important biochemical networks.

Networks in the sense I will use in this chapter can also involve inter-actions between different levels. As we will see, this leads to concepts like downward causation, circular causation and multi-level interactions of all kinds. Surprising as it may seem, the lowest, molecular, levels are controlled by the higher levels. Even DNA is controlled by the organism as a whole.

At first sight biological networks, formed from nucleotides, proteins, membranes, metabolites and many other components, seem fiendishly complex. The DNA sequences that form templates ('genes') for proteins number around 25,000. The number of possible interaction circuits that could be formed by this number is enormous, around $10^{70,000}$.[2] This is not simply astronomical as a number, it is *super*-astronomical. As we saw in Chapter 1, there are 'only' about 10^{80} 'atoms' (actually mostly as elementary particles, such as protons) in the whole known universe. It would take many of the pages of this book to write the number $10^{70,000}$ down in ordinary decimal notation.

We can therefore be certain that the evolutionary process cannot possibly have explored and used more than a very tiny fraction of these possible interactions.[3] If we extend the number of components to include all the other types of chemicals and structures involved, we just make the calculation even more impossible to contemplate. No one, not even the best mathematicians in the world, provided with the largest possible computers in the world, could possibly compute such complexity starting with the molecular components alone. Massive bottom-up reconstruction of living systems is a metaphysical reductionist's pipedream, not real science. Yet many biological scientists still think that it is the way forward. It is more like the search for a needle in a gigantic haystack, comparable in size to the universe.

In order not to be misunderstood, I want to emphasise yet again that we need the reductionist approach to biology. This is emphatically not a book deriding the immense achievements of the reductionist approach. On the contrary, this approach has successfully drilled down to identify the smallest molecular components and their molecular-level interactions. But there is a big difference between acknowledging this great success of molecular biology and adopting it as a method, on its own, for unravelling biological complexity. Pure reductionism will not work, precisely because it does not analyse the kind of complexity organisms display. And, indeed, that is not how scientists like me attempt in practice to understand living systems. We use insights from experiments and theories at all the levels of organisms that we discussed in Chapter 2.

We even use insights from work on the universe itself and from the weather systems on Earth. How do we do that, and can we show that complexity itself is not really an impenetrable mystery? Can we show that it is rather a fundamental property of the universe itself? That is surprisingly easy. Many complex structures form quite naturally.

How Do Complex Structures Form?[4]

Imagine a universe in which all matter is completely uniformly distributed. Perhaps it was like that when the universe was forming from the 'big bang' over 13 billion years ago. Clearly this is not the universe we now know. As we saw in Chapter 1, it is full of strange objects: circles, spirals, horseshoes, vast dust clouds of various shapes and sizes, and of course the even stranger objects in modern theories of physics, such as black holes, dark matter and dark energy. How did such complexity arise from what may have been uniformity and from a perhaps infinitesimally small beginning?

Incidentally, the big bang theory does not imply something from nothing. The idea is just that all the matter and energy of the universe *we can observe* was once concentrated in an extremely small space, represented mathematically as what is called a singularity. What happened at the singularity or 'before' that we simply don't know, nor do we know anything about structures that may exist beyond what we can observe since they would occupy a region of space-time beyond the possibility of their light reaching us. We are not even sure that, or how, time can or should be used in that sentence. All we can say with any degree of assuredness (not the same as certainty) is that we don't and perhaps can't ever know anything about that. We can speculate of course, and many metaphysicians and theoretical physicists do that in theories of multiple universes for example. But those deeper questions do not need to concern us here.[5]

One answer to the origin of complexity is surprisingly simple. There are forces between objects. Some forces – like gravity and the opposite poles of magnets – attract, while others – like the positive poles of two magnets – repel. If we could turn the clock back and distribute the matter of the universe evenly, the particles would immediately start their dance of attraction and repulsion. As they do so they inevitably form networks of interactions. No particle would initially be in a privileged position, but as they attract each other they would congregate to form clumps. Once that happens we break the symmetry of a perfectly uniform universe. Those clumps would form initially as clouds and then as stars and planets.[6]

Breaking symmetry in an unstable system is easy. Small chance events can do the trick. Imagine a ball placed exactly at the top of a hill with a shallow enough top for the ball to have the possibility of staying put. It might stay there indefinitely if there were no chance perturbations. But the slightest wind would displace it from the peak and it would start to roll downhill, all the way if it encounters no insuperable obstacles. Depending on how fine the slope is at the top, it may initially move extremely slowly, but then more rapidly as it experiences steeper slopes. On the way that rolling ball may trigger many other events, such as landslides, that may in turn kill unsuspecting climbers, in turn disturbing their family and friendship structures... the list is endless. Once symmetry has broken, further events can occur simply because of the energy and matter gradients that form, and the chance encounter with other events and situations. There is a continual 'becoming', described by some philosophers as conditioned arising. These are processes with ever more possibilities arising, because each arising forms the conditions for many others. Conditioned arising is a key feature of Biological Relativity, as we will see in Chapter 6.

Depending on the nature of the landscape at each stage, subsequent movement will be predictable only with indefinitely large accuracy in the constants, and in the initial and boundary conditions which specify the landscape. We will encounter the use of the landscape concept in Biological Relativity in Chapters 6 and 9.

We can see this kind of process at work in the weather systems of the sky above us. On a perfectly clear day the sky looks uniform as the apparently evenly distributed particles in the atmosphere scatter sunlight to form the uniform blue colour we see. But the stillness and uniformity mislead us. All the time the atmosphere is exchanging heat, water and gases with the oceans and continents. Convection currents arise as warmer parts rise and colder parts fall. Water particles accumulate. The interactions between them form the wide diversity of cloud structures, from relatively simple smooth planes, to the fiendish complexity of a tornado. No one thing 'makes' the tornado. It makes itself. The forces of matter under the right conditions ensure that these structures

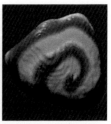

Figure 3.4 Spiral waves in nature at very different scales: galaxy (left), cyclone (middle) and heart arrhythmia (right). A spiral galaxy extends across tens of thousands of light years; a spiral weather system extends across hundreds of miles; a spiral wave in the heart extends across a few centimetres (sources: left: the pinwheel galaxy Messier 101 (www.spacetelescope.org/images/heic0602a, European Space Agency and NASA); middle: http://visibleearth.nasa.gov, Jacques Descloitres, MODIS Rapid Response Team, NASA/GSFC; right: S. Panfilov/Univ.Ghent). For a colour version of this figure, please see the plate section.

should develop. As the complexity increases, so does movement within them. They spin rapidly and rhythmically like tops. The movement in the strongest tornados is so strong that they create immense damage as they hit land and dissipate their energy in a frenzy of destruction.

From a distance above the Earth they appear as spirals. So do spiral galaxies in the depths of space. These also rotate. Our whole galaxy, the one that forms the Milky Way, is rotating, so we are rotating with it, probably around a black hole at the centre of the galaxy. There are rotating structures everywhere in space as well as on the Earth.

We don't need any abstruse kind of theory to explain these formations, both celestial and cosmic. The equations of Newtonian motion suffice, although relativistic effects must also be involved. A similar process creates these rotating structures in both cases. We call them, and many other self-sustaining structures of networks, 'attractors'. The system tends towards these attractors, which explains the name.[7] Note also that no particular part of the network is *the* cause of the attractor. The spiral and its circular motion are properties of the whole network of interactions. They attract more matter and energy to themselves. They are states of the network that attract other parts of the network until the whole network dances to the tune of the attractor. This is how a tornado grows before its accumulated energy finally becomes dissipated in the destruction that it creates on land (Figure 3.4).

Biological Oscillators and Attractors

The structures of the body form attractors in much the same way. Remarkably, the structure and behaviour of a spiral galaxy, a spiral weather system forming a tornado, and what we call a re-entrant arrhythmia in the heart look very similar, although they are occurring on immensely different spatial and temporal scales. To a single cell of the body a spiral arrhythmia in the heart would look as immense as a spiral weather formation seems to us. And like the objects hit by a tornado, those heart cells will receive an abnormal battering from the unusual electrical signals that hit them while the arrhythmia persists. An arrhythmia of this type is a kind of electric thunderstorm in the heart.

Oscillators occur also in single cells, which in turn must look enormous from the viewpoint of a single atom. And, as we know, the components of atoms, their nuclei and electrons, are also spinning around. Oscillators are a ubiquitous feature of the world around us, and we find them at all scales.

We do not need to look, therefore, for a further *general* explanation for why oscillators and attractors form in networks of interactions. Networks whose components attract and repel will automatically form these structures. Where it gets difficult is explaining *particular* examples of oscillators. For example, why do all cells in the body oscillate with daily rhythm, called circadian rhythm? Why do hearts beat rhythmically at or around once per second? How do they come to have such perfect timing, appropriate to the functions they serve? It is when the question of function arises that we need to probe deeper into what is going on. Something must be constraining these oscillators to perform in a way that fits the 'needs' of the organism. Can we find what those constraints are and how they arise? Those constraints are the key to understanding the goal-directedness of living systems.

I put 'needs' in inverted commas because that reminds us that it is easy to get trapped in problems of language in describing what we find in nature. Such words can be seen as shorthand for a more long-winded way of putting the question, which is how do organisms come to be adapted to the environment in which they find themselves? There must be a process of interaction between the environment and the networks of the organism to make this possible. To the extent to which there is a 'need', what we may call a 'meaning' or purpose to what evolves, it arises from this process of interaction. As I will explain further in Chapter 9, it is the contextual logic of a system that provides its meaning. People who tell us that there is no meaning or purpose in the universe are wrong. They can only maintain this view by atomising the universe, as though its components and their behaviour can be seen as isolated. But the facts are the other way round. Nothing is isolated. Even what we call fundamental particles are themselves also waves with extension and interaction. Even at the 'bottom' level, therefore, there are no isolated systems. Biological systems are no exception. They are all open systems.

To explore these questions further, we will look at circadian rhythm and cardiac rhythm as examples and see how they can be explained in terms of interactions between some of the components we discussed in Chapter 2.

Circadian Rhythm

The dynamic networks of cells must have been attracted towards daily oscillators very early in evolution. The daily variation in the energy received from the sun would influence the energy supply, either directly in the case of organisms that can absorb it as plants do, or indirectly by controlling the food supply from other organisms. At the least, therefore, networks metabolising energy would show a daily rhythm. In this case there is a driver of the rhythm: the sun. Every day the metabolic networks would work overtime. At night they would down tools and go dormant. Turning off the energy supply is sufficient to make that happen. The real oscillator here is the rotation of the Earth. Organisms sensitive to the rays of the sun then simply follow that oscillation. They don't generate it themselves.

This is a possible simple explanation for how circadian rhythms began. But we know that the circadian rhythm of cells today is not directly driven in this way. Cells or organisms kept in constant light or dark conditions continue to display daily rhythm. We humans show that phenomenon also when we travel across time zones. Our body clocks continue to function as though the daily variation is that of our place of origin, not our destination. Slowly the body clock will adapt to the new light cycle, and that tells us that there is still some influence of the original driver. But it is no longer the maintaining influence.

What has happened between those early days of the sun being the direct driver and the state of cells and organisms today, where the sun is more like a regulator than a driver? The general answer to that question is that the networks themselves have been attracted towards states in which they automatically display a roughly 24-hour rhythm. The oscillator has become inbuilt. We don't know very much about how that inbuilt property evolved, but we can describe and analyse its mechanism in organisms today in very fine detail.

In vertebrates, what neuroscientists call the suprachiasmatic nucleus (SCN) is a group of cells in the brain that are necessary for circadian rhythm. It contains around 200,000 cells. It has been shown with spectacular success that the fundamental mechanism inside the cells of the SCN is molecular. The feedback loops involve daily rhythms in gene expression level, generated by further feedback loops involving particular protein and gene components. Even single cells isolated from within the SCN can show these circadian rhythms in gene expression.[8]

Mutations in a single gene (now called the *Period* gene) are sufficient to change the circadian period of fruit flies. The discovery of this first 'clock gene' in 1970 was a landmark since it was the first time that a single gene had been identified as playing such a key role in a high-level biological rhythm. At first sight this, and similar discoveries where knocking out a single gene has such a dramatic effect on function, may seem to contradict the view that genes must always act in co-operation to generate higher-level functions. Is that really so? There are various ways of answering this question.

Regulatory mechanisms are often complicated, with many different ways in which they can go wrong. So if a single key component is damaged, and is not backed up in some alternative network, the whole mechanism may malfunction, perhaps in ways that are unpredictable. Most mutations, after all, are a form of damage. Accordingly, when the absence or aberrance of one component is shown to stop a system working, we have to be careful how we interpret this. When a child tinkers with a toy until it ceases to work, it is not necessarily the last thing she did that broke it. The more adult version of this scenario is when a single bad connection in an electrical circuit causes a whole machine, such as a computer, to fail, perhaps because we neglected to make the connections sufficiently

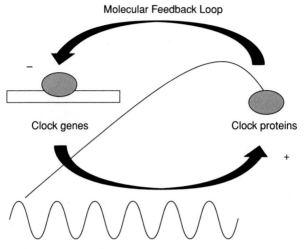

Figure 3.5 The molecular feedback process in circadian rhythm. Transcription and translation of genes called 'clock genes' results in the production of clock proteins, which assemble in the cytoplasm of the cell outside the nucleus. These complexes then move into the nucleus, and further transcription is inhibited. These protein complexes are then degraded, and the clock genes are once more free to undergo transcription. The rates of transcription, translation, protein complex assembly, movement into the nucleus, transcriptional inhibition and protein degradation all combine to generate a 24-hour oscillation (from Foster & Kreitzman 2014).[9]

secure. It is obvious in these examples that the fault or damage is not the sole cause of what is happening.

In biology, some gene mutation and knockout experiments are like that. The results are difficult to interpret. That said, the case of the *Period* gene in the fruit fly is interestingly different. The expression levels of this gene are clearly an integral part of the rhythm generator. They vary (in a daily cycle) in advance of the variations in the protein for which they form a template (Figure 3.5).

But more than that is going on here. The protein is involved in a negative feedback loop with the gene that forms its template. The idea is very simple. The protein levels build up in the cell as the *Period* gene is used as a template to produce more protein. The protein then diffuses into the nucleus where it inhibits further production of itself by binding to what we call the promoter part of the DNA sequence. With a time delay,

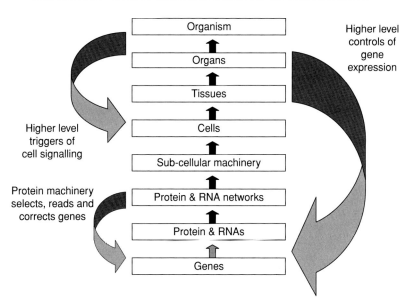

Figure 3.6 Downward causation refers to the control of lower-level processes by higher-level processes. Interaction between all levels can feature this kind of causation. They include the control of cell activity through higher-level processes that lead to hormones, transmitters and other substances acting on the protein receptors in cells; control of gene expression by all levels; and the control of selection, transcription and other modifications of genes and their products, proteins and RNAs.

the protein production falls off and the inhibition is removed so that the whole cycle can start again. So, we not only have a single gene capable of regulating the biological clockwork that generates circadian rhythm, it is itself a key component in the feedback loop that forms the rhythm generator (Figure 3.6).

We call such networks feedback loops because they form self-contained regulatory processes. Feedback is involved in the great majority of cases where downward and upward causation are linked together in biological systems. The linkage is a defining feature of feedback since the downward causation modifies the components responsible for the upward causation, which in turn modifies those generating downward causation . . . and so on. Mathematical models of such loops are easy to construct and they are robust and explanatory. In some cases just two coupled differential equations will do the trick.

Although the basic rhythm generator in this case does seem to be dependent on a single gene and the protein for which it forms a template, we still need to answer the question of whether it carries out its work in isolation? Is it a 'single gene module'? The answer is a resounding 'no'. The further researchers get in unravelling the molecular feedbacks involved in circadian rhythms, the more gene and protein components appear to be involved. At the last count, at least 15 genes are involved.

This has emerged clearly from studies of circadian mechanisms in other animals such as the mouse. Moreover, these rhythmic mechanisms do not work in isolation. There has to be some connection with light-sensitive receptors (including the eyes). Only then will the mechanism lock on to a proper 24-hour cycle rather than free-running at, say, 23 or 25 hours.

It is impressive that a particular gene can be regulated by negative feedback from the very same protein for which it codes. Nonetheless, we don't actually have here an example of a 'single gene' function. But there is a more important point still. Suppose that the original simple feedback between this *Period* gene and the protein it is a template for had indeed been all that was required. Even then it would still not have qualified as a 'single gene function'. The reasons are important in relation to the theory that there is no privileged level of causation.

First, the protein that arose from the template formed by *Period* is like all molecular components – it operates only within the context of the complete cell. It depends for its production on the transcription/translation mechanisms and the ribosome machinery. Its ability to access the promoter region of the *Period* gene depends on properties of the nuclear membrane.

It is not only genes that never operate outside the cellular context. The same applies to individual proteins and the networks they take part in. It is conceptually convenient to 'isolate' the *Period*–protein system. This does indeed help us to appreciate its unusual characteristics as a molecular-level oscillator. But this is an artificial conceptualisation. The real living system operates only in the context of the functioning of many other genes and proteins, and of structures for which there are no DNA templates.

Second, why do we call this the *Period* gene? Because it is a template for the protein that, in feedback with the gene, generates a periodic function. That certainly is the first function of this gene that we have identified. But how do we know what other functions it is involved in? When a part of one system plays a role in another system, the two systems are linked and can be regarded as just one system. Perturbations of one will necessarily influence the other. There may be varying degrees of modularity, by which we mean that some systems are relatively independent of many others. I will discuss an example of that when we look at heart rhythm.

But more often than not, what we perceive to be modularity is part of our way of 'breaking the system up' to make it easier to analyse. That is a perfectly good reductionist stance for analytical purposes, but we need to be aware that it is not more than a convenient stance. Nature will not necessarily conform to our perceptions of her. The *Period* gene has in fact been found to be implicated in embryonic development as the adult fly is formed over several days. And it is deeply involved in the male love songs generated by wing-beat oscillations which are specific to each of around 5000 species of fruit fly and ensure that courtship is with the right species. It would be just as correct, therefore, to call *Period* a development gene or a courtship gene.[10]

Period is therefore rather like one of those very useful pieces of Lego which enable a child to build many different kinds of structure. All 'genes' are like that. Even when we label them with a particular functional name, they are almost certainly playing a role in many other functions too. As we have seen, many other genes are also involved in circadian rhythm. Each of them will also be involved in other functions.

Cardiac Rhythm

Are all biological rhythms based on these kinds of gene–protein–cell-structure interactions? The answer to that question is: yes. Networks necessarily include the structural components of cells. Must the network always include DNA as part of the oscillator? The answer to that question is 'no', not always, and the reasons are both interesting and important. Cardiac rhythm is a good example of a network that includes DNA only as a source of protein templates, not as an integral part of the oscillation network. If proteins were not degraded and needing replenishment, the oscillation could continue indefinitely with no involvement of DNA.

The basic rhythmic mechanism is generated by a relatively compact and tight-knit network of proteins, while the genes that form templates for them are used only to replenish the supply of proteins.

This network functions in the context of what is happening elsewhere in the body – the activity of many other proteins. But in that context this little modular system is sufficient to produce one of the most important oscillators in the body.

Let's have a look at how the oscillator works in this case. The causal system involved works both ways: up and down. The components alter the behaviour of the system, and then the system in turn alters the behaviour of the components. The system is the cell – a muscle cell in the heart. The components are the protein molecules that channel the electrically charged ions (of, in this case, first potassium, then calcium and finally a mixture of elements).

The rhythmic behaviour is shown by the voltage or electrical potential. As the heart beats, the voltage in the relevant muscle cells goes up and down. So does the flow of ions through the protein channels. These two patterns of oscillation are clearly linked. How?

The operation of the protein channels is essential for the critical rhythmic activity of the cell. There is no voltage change unless ions flow through the channels. That is obvious, and that is where we start. But it works the other way round too. The rhythmic activity of the cell drives the operation of the protein channels. We can show this by turning off the feedback from the cellular rhythms to the proteins that channel the ions. If we do that, the system as a whole ceases to function. Neither at the cellular nor at the molecular level is there any longer any sign of oscillation.

Internally within the protein channels, nothing has changed. So if this biological phenomenon were produced exclusively by bottom-up causation, there is no reason why it should stop at this point. But it does. Clearly, therefore, downward causation, i.e. the feedback effects from the system to the components, is essential for the system to function (Figure 3.7).

In Figure 3.7 the top trace shows how the cell voltage varies with time. The bottom traces show how three of the protein channel mechanisms vary with time. These include a potassium channel, a calcium channel and a channel that carries a mixture of ions. There are more proteins involved in the model than I show here, but if I included all of them the

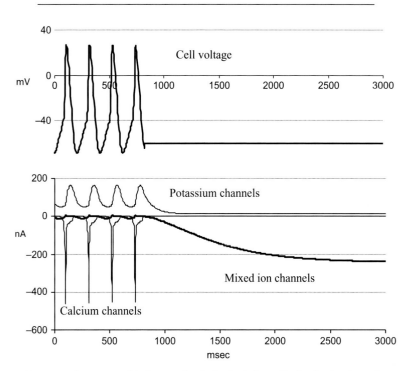

Figure 3.7 Computer model of pacemaker rhythm in the heart. For the first four beats the model is allowed to run normally and generates rhythm very similar to a real heart. Then the feedback from cell voltage to protein channels is interrupted. All the protein channel oscillations then cease. The activities of the channels slowly change to steady constant values. The same experiment can be performed on real cells with the same result (from Noble & Noble, 1984).[11]

figure would become very confusing. The horizontal axis is measured in milliseconds. The vertical axis shows millivolts for the voltage trace and nanoamps for the current traces, representing the activities of the protein channels.

In the first second (1000 msec) there are four oscillations of cell voltage, and corresponding oscillations of protein channels. After four oscillations, the feedback from the cell voltage to the protein channels is cut by holding the cell voltage constant. If one or more of these oscillations of the protein channels were driving the cell voltage, then they would continue to oscillate on their own. But this is not the case. The oscillations of the protein channels cease. In each case, the line that shows their levels of activity flattens out. Clearly, the feedback from the cell voltage to the protein channels is an integral part of the rhythm generator.

This feedback is called the Hodgkin cycle and will be explored further in Chapter 6.

This example shows that, unlike the circadian oscillator, the components of the attractor forming the cardiac oscillator do not include direct involvement of DNA. Moreover, the circadian oscillator can also operate in this DNA-independent mode. Knocking out the 'clock' gene in mice does not stop circadian rhythm. So, the circadian oscillator system includes both types of oscillator, with and without direct involvement of DNA. The cardiac system also includes multiple oscillator mechanisms,[12] but in all of them DNA is simply a template for the replenishment of the protein components.

Two important conclusions arise from these findings. First, important oscillators, necessary for life to continue, are well backed up by additional mechanisms that ensure the rhythm is robust. It can continue to function even when a major component of one mechanism is missing. The second conclusion is that DNA is necessary for the production of the proteins involved, but not always, or even usually, for the function itself.

Functional networks can therefore float free, as it were, of their DNA databases. Those databases are then used to replenish the set of proteins as they become degraded. That raises several more important questions. Which evolved first: the networks or the genomes? As we have seen, attractors, including oscillators, form naturally within networks of interacting components, even if these networks start off relatively uniform and unstructured. There is no DNA, or any equivalent, for a spiral galaxy or for a tornado. It is very likely, therefore, that networks of some kinds evolved first. They could have done so even before the evolution of DNA. Those networks could have existed by using RNA as the catalysts. Many people think there was an RNA world before the DNA–protein world. And before that? No one knows, but perhaps the first networks were without catalysts and so very slow. Catalysts speed-up reactions. They are not essential for the reaction to occur. Without catalysts, however, the processes would occur extremely slowly. It seems likely that the earliest forms of life did have very slow networks, and also likely that the earliest catalysts would have been in the rocks of the Earth. Some of the elements of those rocks are now to be found as metal atoms (trace elements) forming important parts of modern enzymes (see Chapter 2, Figure 2.4).

Gene Expression Patterns

The second important conclusion is that the existence of several functional networks serving a particular function requires that the same genome can be used as a template to generate more than one network. Two questions arise: first, how often does that happen? And, second, how does the organism 'know' which to use? The first question is easy, the second is very difficult.

Let's take the easy question first. It can be answered by looking at the cells in our own bodies. They display an astonishing range of types. They can be as different as bone cells and muscle cells, or as liver cells and nerve cells. As we saw in Chapter 2, a bone cell lays down calcium carbonate as a solid structure, while a muscle cell uses calcium as a mobile ion to control rapid movements of large numbers of contractile proteins to produce powerful movements. A liver cell specialises in metabolic transformations, while a nerve cell develops long, fine tubes called dendrites and axons to enable rapid communication between different parts of the body. That's four very different types of cell already with completely different sets of proteins, what we call expression patterns. There are around 200 more cell types in the body, with their own characteristic expression patterns.

We return to the astonishing fact about these very different cell types: they all have exactly the same genome,[13] derived from the same fertilised egg cell. The answer to the first question, therefore, is that it is very usual for the same genome to be used for very different functional networks as well as for multiple networks serving the same function to make it robust. Our own bodies are very good examples of that. So the answer to the first question is that organisms use the same genome to establish and maintain multiple forms of networks in different cells of the body.[14]

But nature has even more spectacular examples. The brilliantly coloured butterfly gracing your garden with its agile flight amongst the perfumed flowers was once an undistinguished caterpillar crawling up the stems of those plants. They also developed from exactly the same genome. In this case the whole organism displays completely different expression patterns responsible for very different phenotypes at different stages of its life cycle. It is as though the genome contains the templates for two completely different organisms.

Some organisms even switch during their life cycles between unicellular and multicellular forms of existence, as described in the case of the
amoeba *Dictyostelium* in Chapter 2. In a sense we all do. Multicellular
organisms revert to unicellular (the combined egg and sperm) before the
development of the embryo.

What, then, is determining what? Does the genome uniquely specify
the organism, which is an idea that many people think is so obvious that
they take it for granted? Or could it be that things are really the other
way round, or even a combination of the two? Perhaps genes are followers rather than leaders in the evolutionary process. But that would be to
stand standard evolutionary theory, as many of us were taught at school,
right on its head. This book does exactly that, which is why I will discuss
this particular question in more detail in the context of evolutionary theory in Chapter 8. But we can't demonstrate that fully and rigorously until
we have formulated the theory of Biological Relativity in Chapter 6.

The difficult question I left hanging in the air is 'how does the organism "know" which network, which expression pattern, to develop and
use?'. This is one of the central questions in studies of embryonic development. The single fertilised egg cell with, apparently, little structure
divides to form many cells within which, quite early, a polarity is evident: there is a tail end and a head end that can form the context within
which the rest of the branching out into different tissues, organs and
cell types occurs. I wrote 'apparently' because polarity actually exists
right from the beginning, even in the unfertilised egg cell. The fertil-
ising sperm can detect this polarity since it is easier for it to enter at
one pole rather than the other. Symmetry therefore breaks very early in
development. As we have seen, that is not difficult to explain. Symmetry
breaking is a natural and inevitable event in the universe. What is difficult
to explain is the remarkable diversity that then develops in multicellular
organisms.

Not only is that diversity difficult to explain. The diversity amongst
different species is also still difficult to explain. We humans share nearly
all our genome with that of a mouse. With just a few differences, virtually
all the same proteins are made using the same genome templates. That
is even more true when comparing us with a monkey species. And even
comparing humans with a simple worm, like *C. elegans*, we find that we
have similar numbers of 'genes' (as protein templates), which is between

20,000 and 25,000. There are plants that have more 'genes' than we do. It could be that many of the important factors responsible for the differences lie in the regions of the genome that were once labelled as 'junk' – meaning without function – which accounts for more than 95% of the human genome. That idea of 'junk' DNA has now itself been junked as we have found that the great majority of the sequences are transcribed into RNAs. Many RNAs have function within the organism, for example in controlling the genome, and some of them are also inherited independently of DNA. This takes us into areas of modern biological science that are at the very frontier of current research. Epigenetic control of the genome will be discussed in Chapter 8.

Synchronisation of Oscillators: Brainwaves

Earlier in this chapter I referred to the roughly 200,000 cells in the suprachiasmatic nucleus of the brain that are responsible for circadian rhythm. The main oscillator of the heart, the sinus node, also has a similar number of cells. Rhythms like cardiac and circadian rhythms occur also in single cells, so why does nature have so many? When we isolate those cells in physiological experiments they all display rhythm, but the rhythm frequency varies from cell to cell. There is stochasticity at the heart of biological systems. Stochasticity itself is important and we will return to that later in the chapter. First we must ask the question: if all the cells are different, how come that they all beat to the same rhythm in the heart itself? We can ask the same question about oscillators elsewhere in the body.

As with rotating spirals in the universe, the answer is simple. Synchronisation is a natural consequence of interactions in networks. This can be demonstrated in many ways. I will describe just two here, one from an inanimate system, the other from a living system.

Set up a few dozen metronomes on a table and start them oscillating independently and completely asynchronously. What happens? Provided that the table is not completely rigid, the oscillations gradually synchronise all by themselves. There are examples of this kind of experiment in videos on sites like YouTube, so you can verify this remarkable fact yourself. They don't just synchronise approximately. They really do all come together perfectly naturally and with no outside interference.[15] Look carefully and you may notice the tiny vibrations in the table as the synchronisation develops. But even if you can't see the vibrations of the table, you can be sure they exist. The physical properties of the table itself are allowing the individual oscillators to communicate their vibrations to each other and so act as a complete system. The more the table can vibrate the faster the synchronisation. This is a good example of an emergent property. An initial disorder becomes highly ordered through the interactions.

The same kind of mechanism works in the sinus node. About 20 years ago, when computers became powerful enough to do the calculations, I collaborated with two American scientists to perform a similar experiment on the computer model of heart rhythm.[16] We started the calculations with model cells that had a range of around twofold in their natural frequencies. Then we gradually increased the electrical connections between them. In the real sinus node, these connections are formed by special proteins called connexins. Just a very few such connections between the individual cells was sufficient to completely synchronise the whole network. The real connections in the heart itself are stronger than was needed, so there is no puzzle at all as to why a sinus node beats as a single network structure. The same kinds of connections between cells also ensure that this single rhythm is communicated to the rest of the heart.

Coupling between cells can take many different forms in addition to those mediated by connexins. Nerve cells in the brain are connected as networks via tiny junctions called synapses, where chemicals released by one cell and received by another form the links. These kinds of connections can be more specific in their functions than simple electrical connections, since different chemicals can be used at different types of junction. Electrical connections are sometimes found in nervous systems, but the brain would not work properly, or at all, if all the connections were electrical.

Connections to form networks in the brain must be responsible for the various kinds of rhythms, such as alpha waves, that we can record from the whole brain as the electrical signal called the electroencephalogram (EEG). 'Brainwaves' are not just a metaphor! The precise function of such synchronisation in the nervous system is still a hotly debated topic in neuroscience.[17]

Chance at the Heart of the Cell[18]

We come now to an absolutely fundamental question about the nature of biological science. Is it really different from the physical sciences? If so, in what way are they different? The great physicist and developer of quantum mechanics, Erwin Schrödinger, thought that they were and that he could identify precisely the way in which they differ. He gave a series of lectures in Dublin during the Second World War after he had left Germany, disgusted with the racist aspects of the Nazi regime. The lectures were published as a book, famously called *What is Life?*. It is a remarkable book. Published well before the discovery of DNA, he essentially predicted its discovery. He realised that the inherited material containing the templates for the molecules of living systems would need to be what he called an aperiodic crystal. What he meant by that was a structure that repeats itself (that is the 'crystal' aspect of his phrase) as a polymer but that the sequence would have to be different (hence 'aperiodic') in different parts of the polymer in order to be responsible for transmitting so many different inheritable characteristics of an organism.

At that time, 1944, most scientists would have expected this structure to be found in proteins, which is what he himself thought. Instead, it was found in DNA sequences, which have precisely the property of an aperiodic crystal. The same elements, the nucleotides C, G, A and T, repeat themselves, but most often in a different order. That is what enables DNA to form so many different templates for protein and RNA structures. I like the term 'aperiodic crystal' and it conveys an important message that will be revisited several times in this book, which is that DNA is essentially another molecule in the body, just as proteins, metabolites, lipids and other components are. Schrödinger's term eventually gave way to the modern terms: genome, genetic code, genetic programme and so on. The difference in viewpoint will become clearer when we discuss the language of Neo-Darwinism in Chapter 5.

In this chapter, what concerns us is the question of the nature of physics and the nature of biology. Impressed with the aperiodic crystal idea, Schrödinger came to the conclusion that the difference was that in physics stochasticity (randomness) at a lower level is always hidden at higher levels by averaging. This is what enables us to use the equations of thermodynamics at large scales, so that we do not have to bother with the detailed Newtonian mechanics of every molecule of a system. But if inheritance depends on the transmission of specific sequences in DNA then the behaviour of individual molecules cannot be ignored in the same way in biology.

That is true. Otherwise it would not be possible for single mutations in DNA to be responsible for transmitting diseases like cystic fibrosis, sickle cell anaemia, cardiac arrhythmia and some others that can be shown to be traceable to specific mutations in DNA. But what could not have been clear in 1944 is that this is actually the exception rather than the rule in biology. Most changes in DNA do not reveal themselves as a change in the characteristics of the whole organism for the simple reason that organisms are very good at insulating themselves against most changes at the molecular level. A systematic study of all 6000 'genes' in the yeast organism, for example, revealed that 80% of the knockouts (deletions or damage of particular sequences) are silent in the sense that, under normal physiological conditions, the metabolism, reproduction and other important functions of yeast are not affected.[19]

We see a similar result in humans. The surprising fact about many mutations in humans that might be responsible for diseases is that the ability to predict disease states from genome sequencing is actually very poor.[20] In most cases it is weaker than predicting disease likelihoods from family history and lifestyles. In fact, we get much better predictions from family history and lifestyles than from genome sequencing, and the most spectacular improvements in dealing with diseases like cancer have come from changes in lifestyles, such as reducing the prevalence of smoking. The same is true for longevity. The contribution of heredity found in twin studies is only about 25%.[21] Most of the differences in longevity are attributable to lifestyle.

This fact is fundamental. To return briefly to Schrödinger's disgust with the Nazis, the idea of a super race, based on a perfect inheritance, which was the object of eugenics and which led to the disastrous Holocaust, is a political fantasy. We couldn't succeed in doing it even if the eugenicist ideal was desirable. The reason is that from knowing genomes alone we cannot know the organism as a whole. Nor can we select for one 'good' characteristic without the relevant genes affecting many other characteristics, the consequences of which may be unpredictable and possibly far from 'good'. In fact there are no 'good' and 'bad' genes. There are sequences. They are good or bad only in the context of the whole organism, just as the meanings of words depend on their context.

Conclusions

In Chapter 2 we learnt that organisms have different levels of organisation and that the message of this book is that all levels can contribute to generating biological functions that characterise life. In this chapter we have learnt that networks are a key to understanding biological systems since all organisms that can be regarded as alive possess complex networks. It is through the interactions within those networks that functions emerge naturally. Those networks and their interactions are necessarily multi-level. This insight is a fundamental building block of the theory of Biological Relativity, which we will develop in Chapter 6.[22]

All organisms are formed of cells, whether a single cell or many cells. The cell (not just DNA) is also the minimal kit for inheritance. We are now ready to look at some of the remarkable properties of cells and how they choreograph the reproductive and evolutionary dances of their genomes. This is the subject of the next chapter.

Notes

1 The lock-and-key analogy is useful as a first approximation, but it needs qualifying. Proteins are far from being rigidly fixed three-dimensional structures. Many proteins can have several different conformations dependent on other molecules and the whole environment in which they find themselves. They can also have more than one function, a phenomenon called protein moonlighting. Once again, it is the system that is important, and it can constrain the individual molecules in their behaviour and structure. The significance of this kind of constraint for the theory of Biological Relativity will be explored in Chapter 6.

2 The details of the calculation can be found in Feytmans, E., D. Noble and M. Peitsch (2005) Genome size and numbers of biological functions. *Transactions on Computational Systems Biology* 1:44–49. The first person to identify this kind of problem in biology was Elsasser, who called such numbers 'immense numbers'. He meant any number over 10^{1000}. Elsasser, W.M. (1981) Principles of a new biological theory: a summary. *Journal of Theoretical Biology* 89:131–150.

3 This raises the fascinating question of whether there are limits to evolution or whether we should see it as capable in principle of exploring *any* of the huge numbers of possibilities. For further discussion of this question, see the issue of *Interface Focus* devoted to the question *Are there limits to evolution?* (volume 5, issue 6, December 2015); in particular Dingle, K., S. Schaper and A.A. Louis (2015) The structure of the genotype–phenotype map strongly constrains the evolution of non-coding RNA. *Interface Focus* 5; DOI: 10.1098/rsfs.2015.0053.

4 For an excellent introduction to complexity theory, see chapter 6 of Capra, F. and P. Luisi (2014) *The Systems View of Life* (Cambridge University Press, Cambridge).

5 It is a deep error and misunderstanding of current cosmology to represent big bang theory as explaining how everything came from 'nothing'. It would be better to see big bang theory as describing an unfolding.

6 For an example of a simulation of this process in the universe, see Rees, M. (1999) *Just Six Numbers: The Deep forces that Shape the Universe* (Weidenfeld and Nicolson, London; pp. 109–113).

7 Technically, matter and energy from everywhere within a defined area, called the attractor basin, is attracted into the process. For more on attractors in biology see Capra and Luisi, *The Systems View of Life* (2014), pp. 110–118, 176–180 on spiral anatomical features of organisms.

8 Foster, R. and L. Kreitzman (2005) *Rhythms of Life: The Biological Clocks that Control the Daily Lives of Every Living Thing* (Profile Books, London). See also www.voicesfromoxford.org/video/the-rhythms-of-life/367.

9 Foster, R. and L. Kreitzman (2005) *Rhythms of Life: The Biological Clocks that Control the Daily Lives of Every Living Thing.* Yale University Press.

10 Kyriacou, C.P., M.L., Greenacre, M.G. Ritchie, C. Byrne and J.C. Hall. (1992) Genetic and molecular analysis of the love song preferences of *Drosophila* females. *American Zoology* 32(1):31–39.

11 Noble, D. and S.J. Noble (1984) A model of sino-atrial node electrical activity based on a modification of the DiFrancesco-Noble (1984) equations. *Proceedings of the Royal Society of London B* 222:295–304.

12 Details of the multiple oscillators in the heart pacemaker can be found in Noble, D. (2011) Differential and integral views of genetics in computational systems biology. *Interface Focus* 1:7–15.

13 Actually, some genome reorganisation can occur after birth, and there can be some maternal cells, amongst the cells that develop from the fertilised egg cell, passed through the blood circulation during embryonic development. But these exceptions do not change the fundamental point being made here. Even if no such reorganisation was possible and no maternal cells were transmitted, the result would still hold true.

14 A remarkable example of the extent to which the same genome can be used to generate different phenotypes in man is the pair of identical twins studied by Keul J., H.H. Dickhuth, G. Simon and M. Lehmann (1981) Effect of static and dynamic exercise on heart volume, contractility and left ventricular dimensions. *Circulation Research* 48 (suppl 1):163–170. Body mass, respiratory and circulatory functions were all developed differently in one who trained as a runner and the other who trained as a weightlifter.

15 www.youtube.com/watch?v=kqFc4wriBvE. In this video the table movements are initially imperceptible. They become more clearly visible as the metronomes approach synchronisation after about two minutes. The speed with which the synchronisation emerges depends on the elasticity of the table. A completely rigid table would not allow it to happen, but complete rigidity is a physical impossibility. The table may also have a natural frequency, in which case there will be resonance that could amplify the effects.

16 Winslow, R., A. Kimball, A. Varghese and D. Noble (1993) Simulating cardiac sinus and atrial network dynamics on the Connection Machine. *Physica D* 64:281–298.

17 See Terry, J.R., P. Ritter and A. Daffertshofer (2011) Brain modes: the role of neuronal oscillations in health and disease. *Progress in Biophysics and Molecular Biology* 105:1–4.

18 This is the title of a special issue of a journal, *Progress in Biophysics and Molecular Biology*, volume 110(1), that was devoted to articles concerning this topic in 2012.

19 Hillenmeyer M.E., E. Fung, J. Wildenhain, *et al.* (2008) The chemical genomic portrait of yeast: uncovering a phenotype for all genes. *Science* 320:362–365.

20 See Joyner, M. (2015) Has Neo-Darwinism failed clinical medicine: does systems biology have to? *Progress in Biophysics and Molecular Biology* 117:107–112.

21 See the lecture by Tom Kirkwood at www.youtube.com/watch? v=hRUSkIMMhro. Also see Finch, C.E. and T. Kirkwood (2000) *Chance, Development and Aging* (Oxford University Press, Oxford).

22 There may appear to be an inconsistency between the general principle of Biological Relativity (no privileged level of causation) and the statement that certain levels (networks, cells) are key ones. This issue will be dealt with in Chapter 6, where we will see that the general principle is an a-priori one: we should not *assume* in advance of experiments that any particular level is privileged. But we may find as an empirical fact that some levels are more important than others in particular functions of the organism. That is a discovery made a posteriori.

4

Nature and Origin of Cells

Microbiology's Scarred Revolutionary.
Science in 1997,
on Carl Woese, the discoverer of the third domain of life, archaea

The Medical Histology Class

The characteristic aroma of the preservatives and staining chemicals in the histology class at University College London in 1955 still echoes in my mind when I see a histology slide shown in a scientific presentation. As students we were allocated microscopes to view some of the classical slides showing sections of cells, tissues and organs of the body. We tried to draw what we saw and compared our work with that of the textbooks. Soon we were able to identify the nucleus, the mitochondria and many other components of cells and their connections to each other in tissues. Yet, what we were looking at was not really a cell.

Years later I saw a film of a living unicellular organism, an amoeba. It could hardly have been more different from the two-dimensional sections of dead cells that I had drawn as a student. Nothing stood still. Everything was streaming this way and that as the organism moved around. When it found an object that was sensed (I assumed chemically) to be food the movements became beautifully co-ordinated as two extrusions called pseudopodia (false feet) encircled the object, eventually allowing it to be taken in as a membrane coated vesicle to be digested.[1] This tiny organism had a 'nose': the chemical receptors on its membrane surface. It had 'muscles': in fact formed of protein molecules, some of them very similar to those in our cells, only not organised into separate muscular organs. It clearly had a 'nervous system' to connect the two together, although it had no nerves as we know them. It had a clear goal: to feed itself. As we will see later when we discuss the cell cycle, it knows when and how to reproduce itself in an intricately co-ordinated activity when it makes its genes dance as they and their predecessors have done for at least one billion years.

Living organisms, even the simplest of them, are hives of such purposive activity. Later in this book we will discuss this kind of natural purposiveness. Yet, the amoeba has no brain to coordinate these movements. How, then, do they arise? Could the secret really lie somewhere deep in that nucleus? Could there be a genetic programme directing it all like the conducting of an orchestra? That can't really be so because we also learnt as students that there are functioning cells in our bodies that don't have a nucleus. These are the red cells of our blood. We also experimented with nerves that could still conduct signals to coordinate muscle activity. These nerves had been separated from their cell bodies where the nuclei were to be found and could continue to function for a day or two, much longer than the lifespan of many microorganisms. There are other examples also of cells with their nuclei removed that continue to function well until they need to replenish their stock of proteins.

What is it about cells that enable them to do all this? The answer must lie in the activity of the networks inside cells, but many of them must be entirely outside the nucleus. It begins to look as though the nucleus and the DNA are required more as an inner storehouse that can be raided when more proteins are required, and which must therefore be passed on to later generations to enable them to do the same.

Cells as Carriages

If biochemical networks are the workhorses of life, then cells are their elegant and elaborate carriages. Metaphorically speaking, of course, but this metaphor has a grain or two of truth. For, like carriages, cells enclose, compartmentalise and carry. And, like the internalisation of the engine as horse-drawn carriages became automobiles, the 'workhorse' is also carried.

Cells enclose the biochemical networks of the organism since they provide a cell membrane at the surface that may be used by various protein transporters to select substances from the external variable environment to create and maintain the protected and nearly constant internal environment where all the biochemical actions take place. Cells also compartmentalise because, in organisms called eukaryotes, there is a separate nuclear compartment housing the genome, and there are specialised organelles forming further compartments such as mitochondria, plastids and ribosomes, as we saw in Chapter 3. Finally, cells carry all of these with them. In the case of mobile cells they do so literally.

A moving amoeba, for example, transports the complete set of these structures as it moves around. But even cells that do not move independently nevertheless carry all the structures with them as the organism itself moves and as its cells migrate during embryonic development. Some cells in our immune system continue to have free movement even in the adult to enable them to attack, kill and digest invaders. They are like scavengers cleaning the organs of the body by first moving in the bloodstream to circulate to the tissue and organs and then act just like an amoeba seeking food, but in this case their target is material that is dangerous to the body as a whole.

Cells also have supporting architecture to hold their structure together and roadways to transport things from one part to another. Astonishingly, the structural skeleton and the roadways are formed from the same structures, called microtubules. Minute building blocks of proteins called tubulins literally form tubes that, on the scale of a cell, look like fine threads weaving this way and that. It is the way in which these tubules and even thinner actin microfilaments can move that enables the whole cell to move, as in the amoeba example discussed earlier. These are the motors to enable movements to occur. Movements generated internally arise from movements of the internal microfilaments. But they also form the internal parts of cilia and flagella responsible for movements by their whip-like activity on the surface of the cell (Figures 4.1 and 4.2).

The microtubules are also roadways because substances can be attached to the microtubules to move along them all the way from one part of the cell to another. Special small proteins act as motor proteins that attach to the microtubules to form the mechanism by which the protein can move from one end of the microtubule to the other. A whole variety of other molecules and cell structures can attach themselves to

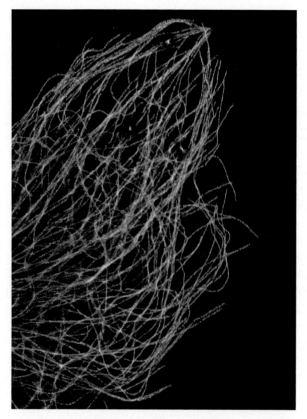

Figure 4.1 Microtubules visualised in a part of a cell. Just as our bodies rely on bones for structural support, cells rely on a cellular skeleton. In addition to helping cells keep their shape, this cytoskeleton transports material within cells, is responsible for cell movement and coordinates cell division. Microtubules are shown here as thin strands. Each tubule is about 24 nm across, but extends over distances that can go from one end of a cell to another, which is many micrometres (from US Department of Health and Human Services (public domain)). For a colour version of this figure, please see the plate section.

the motor proteins to be transported. These can be RNAs, proteins, protein complexes, organelles and whole vesicles. The roadways are two-way. There are motor proteins designed to go one way or the other.

The existence of these roadways in cells solves an otherwise puzzling problem. Cells need to inform their genomes in the nucleus which

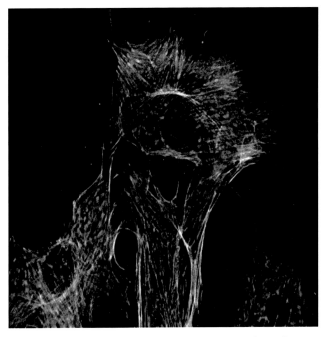

Figure 4.2 Microtubules (green) and actin microfilaments (orange) in an epithelial cell from the lung of a cow. The tubules are about 24 nm across; the filaments about 6 nm across (from US Department of Health and Human Services (public domain)). For a colour version of this figure, please see the plate section.

proteins to make. One way of doing that in a highly specific way is to use the microtubule transport system to carry information on the state of small regions of the cell, for example just beneath a membrane channel protein, all the way to the nucleus. Up and down regulation of genes can therefore be controlled in response to the finest structural and dynamic features of cell activity. This was a puzzling problem since some ions and molecules are used to signal many different aspects of cell function. Calcium is frequently used in this way. The information would not be very useful if it depended only on the global cell calcium. Being able to detect changes specific to small regions of the cell enables the same messenger chemical to signal many different aspects of cell function. This is the way in which the genome 'knows' about fine cellular structure and what is happening dynamically in all parts of that structure.[2] Calcium, and other signalling molecules, are therefore a bit like traffic lights. And, just like traffic lights, what is controlled depends on where the traffic lights are placed.

Finally, microtubules are also the puppet strings that hold the genetic material and move it around in the remarkable 'dance of the genes' that occurs during cell division, as we will see later in this chapter.

The Simplest Cells: Bacteria and Archaea

Not all cells are compartmentalised in this way. Prokaryotes are much simpler and must have evolved before eukaryotes since, as we will see later, eukaryotes arose by symbiotic fusion of prokaryotes. Prokaryotes include bacteria and a group of cells called archaea, meaning 'ancient'.

Originally it was thought that there was only one domain of prokaryotes, the bacteria. One of the great scientific benefits of genome sequencing is that it becomes possible to compare sequences in many different species. We can then trace the history of the sequences and derive a phylogenetic tree showing the most likely way in which each group of organisms is related to the rest. In 1977 Carl Woese used the sequences in ribosomal RNAs to derive the tree shown in Figure 4.3. Later work using DNA sequencing and chemical characterisation of the proteins forming the physiological networks in cells has fully confirmed this interpretation.

Bacteria

Bacterial cells have no nucleus. Their DNA simply floats free in the cytoplasm, usually as a single circular thread. A few have an even simpler linear thread. It is virtually certain that DNA existed in these or even simpler forms long before the evolution of the highly complex structures forming the chromosomes in the nuclei of eukaryotes. The single thread is usually coiled up to form a kind of nucleus lacking a membrane, called a nucleoid, but it opens into a looser form when the cell grows large enough to divide into two. In this open form the DNA is replicated and each daughter cell receives a copy, which then coils up again. As we will see, this particular 'dance of the genes', already complicated enough, is nevertheless remarkably simple compared to the elaborately orchestrated events in eukaryotes.

Phylogenetic Tree of Life

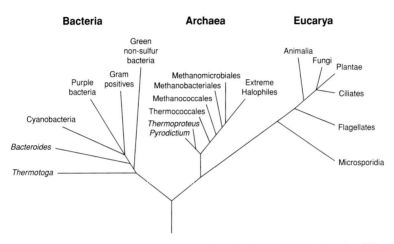

Figure 4.3 Carl Woese's interpretation of the relationships between the three domains of life: bacteria, archaea and eukarya. Woese showed that at a molecular level archaea are more like eukarya than bacteria even though their structure seems to resemble bacteria more than eukarya, so they are represented as branching after the division between bacteria and other forms. Note also that animals, plants and fungi form quite small branches. In terms of biomass, this is a correct representation of the relative importance. From: *Proc. Natl. Acad. Sci. USA* vol. 87, pp. 4576–79, June 1990.

Bacterial reproduction is therefore by simple fission and is asexual. If this were all that happens then variation could only happen by mutations in the cell's own DNA. It may seem therefore that one of the advantages of sexual reproduction, which is the creation of variants by mixing up two different genomes, is not available to bacteria. But in fact, some bacteria at least are highly promiscuous with DNA. They have their own forms of multiple-partner 'sex' and they do it all the time, not just when they reproduce. There is a veritable orgy of DNA sharing amongst organisms at the micro scale, where individuals acquire new characteristics wherever and whenever they can. Those characteristics are then inherited.

Does that kind of inheritance sound familiar as a rather heretical idea? The evidence for and the significance of the inheritance of acquired characteristics will be dealt with further in Chapters 5, 7 and 8. Meanwhile, let us note that the fraction of life on Earth represented by bacteria greatly exceeds that of all the plants and animals. They almost certainly existed for at least one billion years before plants and animals. To anticipate

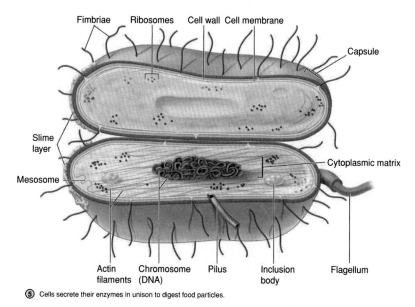

Fimbriae   Ribosomes   Cell wall   Cell membrane   Capsule   Slime layer   Mesosome   Cytoplasmic matrix   Actin filaments   Chromosome (DNA)   Pilus   Inclusion body   Flagellum

⑤ Cells secrete their enzymes in unison to digest food particles.

Figure 4.4 Typical structure of a bacterium. There is no nucleus. The DNA is instead coiled up to form what is called a nucleoid. Various protrusions form structures that can propel the organism (flagella) and transfer DNA between bacteria (pili). A bacterium with this kind of shape (there are many other shapes!) would be about 5–10 μm long and about 1 μm wide (https://commons .wikimedia.org/wiki/File:Major_events_in_mitosis.svg). Copyright McGraw-Hill companies inc. For a colour version of this figure, please see the plate section.

themes taken up in those later chapters: the standard theory of evolution presupposes a separate germline isolated from the rest of the organism, and to which the concept of the Weismann Barrier could apply.[3] Before the evolution of a separate germline the theory would not apply. It could not therefore be relevant to the great majority of the period during which evolution has happened on Earth. Moreover, the very concept of a species does not have much sense at the scale of microorganisms.

Bacteria have no fewer than three ways in which they can acquire new DNA sequences. The first is simply to take in DNA from their environment. The second is transfer of DNA via viruses called bacteriophages. The third is a little more like 'real sex': transfer of DNA directly from cell to cell by conjugation. The mechanism by which they incorporate DNA from their environment, which is called transformation, seems to be an adaptation to facilitate this process since it depends on a large number of proteins in the cell, and it also seems to be an adaptation to stressful

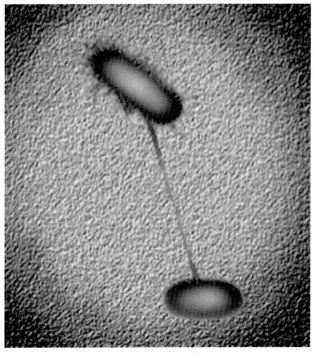

Figure 4.5 Image of the connection between two pili stretching out between the two bacteria. Each bacterium is about 1 μm across. The bridge is about 6 μm long. It contracts to draw the two bacteria close together. Exchange of DNA can then occur (http://image.slidesharecdn.com/). For a colour version of this figure, please see the plate section.

environments. Transfer of DNA is one of the ways in which bacterial populations can develop resistance to antibiotics.[4]

Conjugation is also a useful adaptation and depends on a remarkable form of cell-to-cell connection, first discovered in 1946 by the American scientist Joshua Lederberg and his thesis supervisor Edward Tatum. The cell surfaces of bacteria have protuberances called flagella that enable the bacterium to move by using a whip-like motion. Some of these protuberances, called a pilus, can reach out towards other bacteria to draw them in. The recipient bacterium can then put out its own pilus, joining with the first to draw the bacteria close together. DNA can be transferred between them (Figure 4.5).

Archaea

Woese's identification of the archaea as a different domain depends almost entirely on a molecular approach. At a more superficial cellular structural level, bacteria and archaea have a lot in common, which is why they were not originally thought to be from a different lineage in the evolution of life. But at the molecular level of DNAs, RNAs and proteins there are at least ten major differences. Moreover, in many of these, including some very important proteins, archaea actually resemble eukaryotes more than bacteria.[5] This is the reason why the most recent 'trees of life' place archaea and eukaryotes on a common stem separate from bacteria.

They were also initially thought to be found only in extreme environments, like hot springs and salt lakes. Some live as deep as 3 km underneath the surface of the Earth at a temperature of 85 °C, where they reduce carbon dioxide to produce methane. If global warming really becomes a complete run-away process, they may well be the only organisms left! They have now been found to occupy most environments on Earth. Vast numbers form part of the tiny organisms called plankton in the oceans.

Another great contribution to the study of the prokaryotes was also made by Carl Woese. Archaea share the bacterial propensity for promiscuous sharing of DNA. Horizontal gene transfer has occurred and still occurs frequently amongst prokaryotes, and also occurs to some degree amongst eukaryotes. We will look at the evolutionary significance of this fact in Chapter 7.

Woese was actively sidelined by some of the biological scientific establishment in the twentieth century. The prominent Neo-Darwinist Ernst Mayr objected strongly in 1998 to his division of the prokaryotes into two domains. But by 2003 Woese was honoured by receiving the Crafoord Prize, an equivalent to the Nobel Prize in non-medically oriented biology. The prize was precisely for the discovery of 'the third domain' of life.

The field of evolutionary biology seems to bring out very strong passions in people – positive and negative. It is almost as though the less illumination there is on the question of how evolution actually happened, the greater the heat the debates generate. When it concerns the origins of cellular life, the one thing we can currently be very certain about is that the answers to most questions are still very uncertain.

Eukaryotes: The Largest Organisms but the Smallest Domain

You will have noticed in Figure 4.3 that cells of the kind we have in our bodies, the eukaryotes, seem to form a relatively small part of the network of life. Animals and plants form an even smaller part. Is this really a correct reflection of reality? After all, what we easily see around us are animals, plants and fungi. It was the study of these that gave rise to the great work of Lamarck, Darwin, Wallace and others in establishing the now incontrovertible evidence for the transformation of species. They are also what you will usually see in zoological and botanical gardens, although some science museums are increasingly featuring the strange world of prokaryotes.

Appearances are deceptive. In fact, on two major counts, duration and quantity, the prokaryotes easily dominate life on Earth. They may have done so alone for one billion years or so from the period of origin of cellular life, which may have been around 3.5 billion years ago. There are microfossils from that long ago. The earliest eukaryotic fossils date from 1.7 billion years ago. Of course, they may have existed earlier without leaving a fossil record that we have so far been able to find. The prokaryotes also dominate in quantity. There are vastly more of them than there are of us and our fellow eukaryotic creatures. It is estimated that the total quantity of carbon locked up in prokaryotes is similar to the amount locked up in all plants on Earth.

There is another sense, too, in which we, and the whole domain of eukaryotes, are the smaller domain. We and other animals and plants are made from prokaryotes, and we are still totally dependent on them. The cells in our bodies, like all eukaryotic cells, arose from symbiotic association.

The fusion of certain prokaryotic forms to create the immensely complex organisation of compartmentalised eukaryotic cells was a major step in evolutionary history. The cellular apparatus in eukaryotes is so rich and remarkable that their evolution must have occurred in many stages.

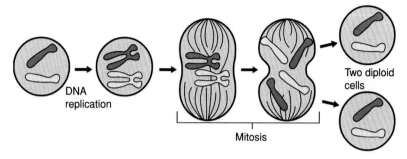

Figure 4.6 Diagram of the stages of nuclear mitosis (www.ncbi.nlm.nih.gov/About/primer/
genetics_cell.html).

There must then have been many stages in the development of multicellular forms of life, finally leading to the great radiation of such forms in what is called the Cambrian explosion, around half a billion years ago. Within a few million years, just a blink of an eye in terms of a history spanning four billion years, all the main multicellular life forms we see today developed in a spectacular radiation.[6]

It is in this context that we can understand another important fact about eukaryotes and prokaryotes. Eukaryotes have added no new metabolic energy transforming proteins. Both respiration and photosynthesis are carried out by mitochondria and plastids respectively, which of course came from symbiogenesis involving prokaryotes. Many of the basic functional processes of life originated in prokaryotes.

We can best appreciate just how great the difference in organisation is by comparing the cell cycles in prokaryotes and eukaryotes (Figure 4.6).

The Cell Cycle

The way in which cells without nuclei – the prokaryotes – divide and reproduce is clearly not possible when the DNA is locked up inside the nucleus. The nucleus is rather like a cell itself – a cell within a cell – and this is almost certainly how it came into being. A fusion of cells into symbiotic union must have created the first eukaryotes from prokaryotes. You might expect therefore that when a eukaryote divides there must be several stages since the cell within the cell also has to divide. The precise details vary, but the sequence is: reproduction of DNA within

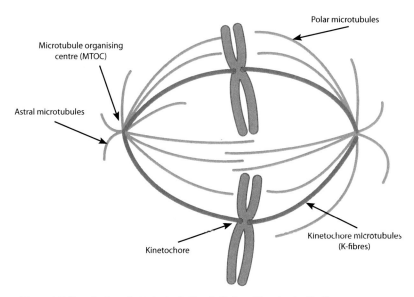

Figure 4.7 Organisation of a typical mitotic spindle found in animal cells. Chromosomes are attached to kinetochore microtubules via a multiprotein complex called the kinetochore. Polar microtubules interdigitate at the spindle midzone and push the spindle poles apart via motor proteins. Astral microtubules anchor the spindle poles to the cell membrane. Microtubule polymerisation is nucleated at the microtubule organising centre (Lordjuppiter (own work) (source: http://commons.wikimedia.org/wiki/File:Spindle_apparatus.svg (CC BY-SA 3.0 (http://creativecommons.org/licenses/by-sa/3.0)))).

the nucleus, division of the chromosomes, division or dissolution of the nuclear membrane, separation of the two sets of chromosomes and finally division of the complete cell.

The first two stages leading to two separate nuclei is the process we call mitosis. The last stage is called cytokinesis. The complete process has to be seen to be believed. A whole army of protein networks are employed by the cell to orchestrate a grand dance of the chromosomes, worthy of a Tchaikovskian ballet. Or, given the overtly sexual connotations, perhaps Stravinsky's *Rite of Spring* (Figure 4.7)?

First, each of the chromosome's DNA threads are replicated. A set of proteins carries out this process and checks for the accuracy of copying, repairing mistakes as required. This process is carried out while the DNA is available to be read. Once the real dance begins, the DNA threads can no longer be copied since they then coil tightly around the histone proteins to form the now visible chromosomes. Those appear as double chromosomes, hitched together at the middle of each pair by yet another protein complex. The next stage is the most remarkable. Each of the chromosome pairs gets attached to filaments that form as a spindle, each half ready to pull the separated chromosomes to opposite ends. The nuclear membrane then divides or, in those cells where the membrane dissolves, the membrane reforms, so that two equivalent nuclei now exist. The cell itself then divides and in the process a nucleus and a share of all the organelles, membranes, microfilaments and free chemicals is allocated to each daughter cell.

The process is clearly theatrically dramatic. The chromosomes are literally like puppets on strings, dancing to the tune of the cell. It takes about an hour, which is short compared to the length of time between mitotic periods. What sets it all in motion? Is there a nuclear 'clock' dictating the timing? No, the trigger is a cytoplasmic protein, one of a set of proteins controlling the cell cycle and therefore called cyclins (Figure 4.8).

It is important to note that in this complex sequence of events, DNA does not directly do anything other than be copied and separated by the sets of proteins that perform these tasks. It does not directly command the process. It is the cell networks that choreograph the ballet. The DNA is a passive participant.

This is also the direction of causality when DNA is being used as a template for making proteins during the long periods between mitosis and cell division. The signals for DNA to be transcribed come from the rest of the cell. During the normal 'resting' phases, the transcription of DNA to form RNAs and proteins is signalled by proteins called transcription factors that attach to regulatory sites on DNA. The levels of the RNAs and proteins formed are also dependent on the degree of methylation of DNA, and on binding of regulatory molecules to the chromatin proteins. There is continuous interaction between the cell physiology and DNA transcription in order to maintain cell function. DNA itself does not interact directly with the protein-forming machinery. It is RNA that performs that role. Only when DNA has been transcribed into RNA does that machinery get activated.

What triggers mitosis? Once again, the answer lies in the cell networks and the proteins forming them. The final signal for the transition to begin mitosis is a protein forming part of a family appropriately called cyclins. It was the identification of this cyclin and a cyclin-dependent kinase that was celebrated with the Nobel Prize awarded in 2001 to the British

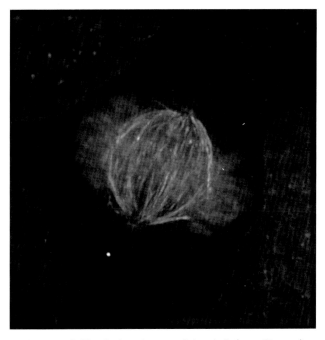

Figure 4.8 Cell in mitosis at the stage of the spindle (green filaments) having formed before the chromosomes (red) are separated and pulled to each end of the cell (Roy van Heesbeen, via Wikimedia Commons). For a colour version of this figure, please see the plate section.

biologists Paul Nurse and Timothy Hunt and the American Leland Hartwell. The genes that form the templates for these proteins are often referred to as the genes that 'control' the cell cycle. In particular, the gene called *cdc2* (**c**ell **d**ivision **c**ycle gene 2 – part of a family of around 100 such genes) was identified by Paul Nurse as the 'start' gene in a particular strain of yeast cells. The discovery was of great importance both for the fundamental nature of cell biology and for research on diseases like cancer, where cells proliferate 'out of control' and invade the rest of the body.[7]

But readers of this book will already suspect that I don't go along with the concept of causality often used in science popularisation to describe what is happening here. A single stretch of DNA cannot 'know' when to trigger mitosis. No single gene, or even a set of genes, can possibly have this kind of natural dynamic purposiveness. That purposive behaviour lies in the networks, since they have the dynamics necessary for the logical operations required to work out when the cell state is ready for division and what to do to ensure that, usually, this takes place accurately. This is natural purposiveness and we will encounter more examples in later chapters. Such purposiveness necessarily emerges at a level that has the degree of dynamic complexity necessary for it to be instantiated. Cells clearly have that complexity. Molecules do not. It goes without saying that this is not conscious purposiveness. It is purposiveness in the sense that the organism responds to challenges in a logical way. We don't need to know how the logical purposiveness developed in order to recognise it in the behaviour of an organism.

Meiosis

The intelligence to choreograph mitosis is remarkable enough. The story gets even better when we consider what happens in the germ cells of organisms that use sexual reproduction.

The functional significance of sex is still a hot topic of debate in biology. As we saw earlier in this chapter, exchanging DNA in prokaryotes could be an effective way for populations to adapt by discovering new combinations of DNA sequences, and this has been a major factor in the evolution of bacteria and archaea. As we will see in Chapter 7, acquiring DNA from other organisms is not absent in eukaryotes but it is thought to be far less frequent. The main process by which mixing of genomes can occur is through sexual reproduction when genome reorganisation and recombinations can occur at two different stages. The first precedes DNA replication. Before the cell divides, exchange of DNA occurs between each of a pair of chromosomes, each of which originally came from one of the two parents. The cell then divides yet again, but there is an important difference. The second division is not preceded by a further replication. Instead, each daughter cell, now called a germ cell or gamete, receives only one of each kind of chromosome. The cell is said to be haploid, instead of diploid (Figure 4.9).

This enables a further round of mixing when the germ cells of two different organisms fuse through the process of fertilisation to restore the double chromosome number.

Once again, all of this is orchestrated by networks incorporating cyclin and related proteins.

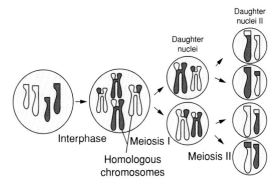

Figure 4.9 Diagram showing the stages during meiosis. This is the equivalent diagram to that for mitosis shown in Figure 4.6 (National Institutes of Health, public domain).

RNA and Other Early Worlds?

As we have seen, the genome and the systems that have evolved for managing it are extraordinarily complex. How could it have evolved and what did it evolve from? These are difficult questions to which there are no secure answers as yet. We might have hoped that comparison of DNA sequences would give us some clues, as indeed it has about the later stages of evolution. But massive lateral exchange of DNA during the early cellular stages when prokaryotes were the only cell forms in existence would have blurred any evidence of very ancient lineages. This is the main reason why all the modern diagrams of trees, networks and circles of life, relating living organisms to each other, fade into uncertainty at the bottom.

But we can be sure that a system as complex as the genome could not have existed in the early stages. DNA alone is not alive. It depends on living systems to perform its functions. This is one of the reasons for thinking that the evolution of RNA and its interactions must have preceded that of DNA. On this view there was an RNA world earlier in evolution. This view also fits with the fact that DNA is not used directly to manufacture proteins; RNAs perform that function.

Conceivably, such worlds may exist elsewhere in the universe, which is an exciting possibility for extra-terrestrial biology. Another fact favouring this interpretation is that RNAs can also act as enzymes, which are necessary to speed up biochemical reactions. Functioning biological networks using organic enzymes could therefore have existed before the evolution of DNA and complex proteins. And before RNA it is likely that inorganic catalysts played that role. The really ancient roots of metabolic networks may lie in hot, rocky crevices containing the essential trace elements (Figure 2.2), acting as natural facilitators of the speed of reactions. Those elements are essential because even today they are necessary as part of the active sites in many protein catalysts. Perhaps those trace elements are a kind of 'living fossil' evidence of the geochemical origin of life. The fact that they are only found in combination with proteins (not nucleotides) suggests that peptides evolved before nucleotides in an early metabolism-led phase of evolution.

What, then, could have formed the membrane channels necessary to control the internal environment? Today, those are all formed by complex proteins, but the highly developed forms of proteins that enable such elaborate channels to occur are very unlikely to have existed before the development of RNA and DNA. The answer is that there are simpler peptide molecules that can form pores in lipid membranes. Some of these, such as valinomycin,[8] can form pores that are highly selective. There is evidence that a limited number of peptide- and protein-forming amino acids may have existed four billion years ago.[9]

The problem remains, what could have served the container function before the evolution of membranes? A suggestion made by Michael Russell working at NASA's Jet Propulsion Laboratory is consistent with the geo-thermal view of the origin of life. This is that iron sulphide could have served the role. The elements iron (Fe) and sulphur (S) are amongst the important trace elements in organisms (Figure 2.2). This is in the 'way out' category of hypothesis, but it is in the very nature of work on the origin of cells that we have to address the question of what could have come before them. At present, that must be sheer speculation. 'Way out' ideas are all we have.[10]

How Cells Form Tissues

Multicellular organisms form tissues, organs and whole systems, as we saw in Chapter 2. Do the cells just congregate, rather as the amoebas do when they form a slime mould? The answer to that question is that more permanent connections have evolved. A family of proteins, appropriately called connexins, form channels that can span across the bilayer membranes of two cells. Those channels allow many molecules to pass that become, as it were, of common ownership within the tissue. They also contribute to its structural stability, although this is also ensured by specialist cells forming what is called connective tissue.

The Nearly Cells: Viruses

Viruses are the tiniest particles of 'living' material, so small that an electron microscope is necessary to visualise them. Most consist of little more than some DNA or RNA enclosed or wrapped around a protein structure. They are not living organisms by the usual definitions. They require a host cell to use its special protein machinery to replicate its DNA. Their evolutionary origin is unclear. They can't be the original living organisms since they require other organisms to survive and be copied. They are clear evidence, if any more were now needed, that DNA alone is dead, inactive.

Although their origins are unclear, their role in evolution may have been very important since they can be a source of lateral transfer of DNA and RNA.

Tree, Networks or Rings of Life?

The idea that species transform into new species, championed by Lamarck, was already a basis for supposing that there must be lineages, perhaps going back to common ancestral forms. Darwin formalised this in his famous tree of life sketch (see Figure 5.1). Modern developments of that sketch now abound. Some continue to represent it as a simple tree: a single trunk with branches but no interconnections between them. That is clearly no longer tenable. We have to recognise that for the great majority of life lateral gene transfer made the tree into a network at its roots at least, but probably elsewhere too. Symbiogenesis to create organelles like mitochondria and chloroplasts make parts of it look like a ring since formerly separate branches reconnected.

For this book I searched for a modern diagram that represents all these new and very significant findings. I found Figure 4.10 in a lovely book, *In*

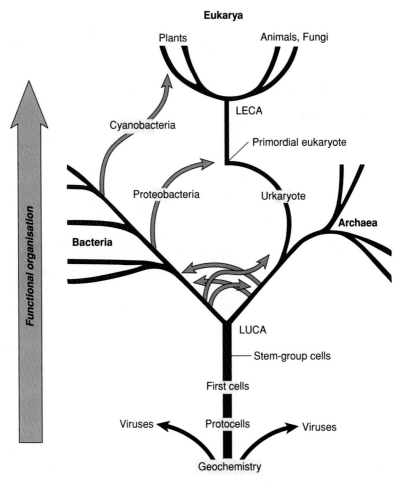

Figure 4.10 A tentative 'tree of life' proposed by Franklin Harold. Viruses and the earliest cells are represented as having arisen from the Earth's early geochemistry. The diagram includes the lateral transfers that create networks of inheritance of DNA, and the 'rings' arising from incorporation of proteobacteria and cyanobacteria into eukaryotes. LUCA is the postulated last universal common ancestor, but it should be understood that this may have been a pool of such ancestors. The concept of separate organisms would have been inappropriate for a period before well-defined cell enclosure by membranes occurred. LECA is the last eukaryotic common ancestor (from Harold 2014).[11]

Search of Cell History, by the American cell biochemist Franklin Harold.[12] Viruses and the earliest cells are represented as having arisen from the Earth's early geochemistry. We know very little indeed about what may have happened during that early period, nor when exactly it happened.

It is quite possible that it happened several times before the evolution of organisms really took off. We do not know how the earliest cells and viruses were related. Viruses might have been protocells that regressed and lost their independent ability to reproduce. A purely virus world would be impossible since viruses depend on using a cell's machinery to reproduce.

It is then thought that there was a common ancestor (LUCA: last universal common ancestor) before the division into bacteria and archaea. Eukaryotes then branched off from archaea to form the LECA (last eukaryotic common ancestor) from which plants, animals and fungi then branched off. There are two important features of this illustration that should be emphasised. First, there was extensive exchange of DNA material during the early period of evolution. This continues, though thought to be at a less frequent rate, in the later stages.[13] Second, the great majority of the evolutionary process occurred before the emergence of animals, plants and fungi. Unicellular organisms are by far the most abundant ones on Earth even today.

The Death of Cells

An important process in multicellular organisms is cell death, or in technical language, apoptosis. This is also a kind of dance initiated by the environment in which the cell exists. The tissues and organs of the body 'know' when some cells are not needed or even when they become dangerous. An example of not being needed would be the cells that exist between what are to become the digits of a limb. The developmental process eliminates them to enable the separate fingers or toes to form. An example of being dangerous would be cancerous cells. Apoptosis is often called 'programmed cell death'. That terminology is okay provided we don't draw the conclusion that there is a cell death programme in our DNA.

You might think that a DNA programme for death, including cell death, must surely exist. Aren't we programmed to die before a certain age to leave room for the young? Well, no, not really. Consider this. Longevity is an extremely rare phenomenon in nature. Wild species generally live much shorter lives compared to what is possible in a sheltered environment. Modern biological studies of the ageing process have shown that the heredity component in ageing is only around 25%, and attempts to find a genetic programme for death have failed despite many attempts in organisms like nematode worms.

Conclusions

This chapter concludes the trio of chapters (2–4) introducing but also radically re-interpreting some of the main features of biology to prepare the reader for the following chapters, where the consequences of a relativistic interpretation of biology will be developed. Many hints have already been given in these early chapters of what is to come. There will therefore be strong echoes of those hints as you read on. As noted in the Preface, there will be a little repetition in some of the chapters, albeit always from a different angle. Novel and perhaps unexpected re-interpretations are best taken in by stages. I have also tried to make each chapter self-contained so that readers who prefer dipping in rather than cover-to-cover reading can quickly get the message of the book.

We are now ready to tackle the big questions of evolutionary biology and of Biological Relativity. Chapter 5 will introduce the orthodox Neo-Darwinist view of evolutionary biology. Chapter 6 will introduce the principles of Biological Relativity. Chapters 7 and 8 will then discuss what needs to replace Neo-Darwinism as an exclusive theory of evolution. The principle of relativity requires an inclusive theory.

Notes

1 A good example is at www.youtube.com/watch?v=AaocvmsD_2Q
2 As an example, see Ma, H., R.D. Groth, S.M. Cohen, *et al.* (2014) γCaMKII shuttles Ca2+/CaM to the nucleus to trigger CREB phosphorylation and gene expression. *Cell* 159:281–294.
3 The Weismann Barrier supposes that the germ line in multicellular organisms is isolated from any changes in the other cells of the body. This idea will be discussed further in Chapter 5.
4 Precisely this process has now been shown to face the world with a potential health disaster. Scientists in China have shown that horizontal DNA transfer between bacteria has allowed antibiotic

resistance to develop in a way that could be the beginning of what some have called the antibiotic apocalypse, a post-antibiotic era plunging the world back into the 'dark ages' of medicine before antibiotics were discovered. Bacteria with the transferred *MCR-1* DNA sequence (gene) were found in both pigs and humans. Liu, Y.Y., Y. Wang, T.R. Walsh, *et al.* (2015). Emergence of plasmid-mediated colistin resistance mechanism MCR-1 in animals and human beings in China: a microbiological and molecular biological study. *The Lancet* 16(2):161–168.

5 Harold, Franklin (2014) *In Search of Cell History* (University of Chicago Press, Chicago, IL). See also Spang, A., J.H. Saw, S.L. Jørgensen, *et al.* (2015) Complex archaea that bridge the gap between prokaryotes and eukaryotes. *Nature*; DOI:10.1038/nature14447: 'the archaeal ancestor of eukaryotes had a dynamic actin cytoskeleton and potentially endo- and/or phagocytic capabilities, which would have facilitated the invagination of the mitochondrial progenitor'.

6 This spectacular radiation was, however, limited to animal phyla. Flowering plants appeared much later. Their evolution involved hybridisations and genome duplications that form part of the natural genetic engineering processes described in Chapter 7. See Davies, T.J., T.G. Barraclough, M.W. Chase, *et al.* (2004) Darwin's abominable mystery: insights from a supertree of the angiosperms. *PNAS* 101:1904–1909.

7 I have not included a section on cancer in this chapter because the initial enthusiasm for thinking that a few simple changes might enable us to understand how cells move into a cancerous state of uncontrolled cell division has given way to the realisation that cancer is a multi-factorial problem that is not yielding to that kind of approach. Weinberg, R.A. (2014) Coming full circle: from endless complexity to simplicity and back again. *Cell* 157:267–271. See also articles in a special issue of *Progress in Biophysics and Molecular Biology*, volume 106; Soto, A., C. Sonnenschein, P.K. Maini and D. Noble (2011) Systems biology and cancer. *Progress in Biophysics and Molecular Biology* 106:337–339: 'while the achievement of sequencing the complete human genome, and those of other species, has been of great benefit to fundamental science, for example in comparative genomics and evolutionary biology, it has not led to the expected quick and simple solutions to multifactorial diseases'.

8   Valinomycin is a ring peptide that is highly selective for potassium ions.
9   Longo, L.M., J. Lee and M. Blaber (2013) Simplified protein design biased for prebiotic amino acids yields a foldable, halophilic protein. *PNAS* 110(6):2135; DOI: 10.1073/pnas.1219530110.
10 For further reading in this fascinating area, see Lane, N. (2015) *The Vital Question: Why is Life the Way It Is?* (Profile Books, London). Nick Lane also favours the metabolism-first approach to the origin of life. Further support for the metabolism-first approach comes from the work of Lee Cronin's group in Glasgow, who have shown that 'peptide bond formation from unactivated amino acids is less challenging than previously imagined. This process is both simple and general, does not require catalysts or activating reagents and can produce large yields of oligomers.' Rodriguez-Garcia, M., A.J. Surman, G.J.T. Cooper, *et al.* (2015) Formation of oligopeptides in high yield under simple programmable conditions. *Nature Communications* 6:8385; DOI:10.1038/ncomms9385.
11 Harold, Franklin (2014) *In Search of Cell History* (University of Chicago Press, Chicago, IL).
12 Harold (2014) *In Search of Cell History.*
13 It was originally thought that such transfer would be impossible in multicellular organisms. But it is important to remember that all multicellular organisms using sexual reproduction go through a unicellular phase, the unicellular egg and sperm. Sperm have been shown to acquire DNA from their environment and transfer this DNA to the offspring. Pittoggi, C., R. Beraldi, I. Sciamanna, *et al.* (2006) Generation of biologically active retro-genes upon interaction of mouse spermatozoa with exogenous DNA. *Molecular Reproduction and Development* 73:1239–1246. 'These results indicate that an efficient machinery is present in mature spermatozoa, which can transcribe, splice, and reverse-transcribe exogenous DNA molecules. This mechanism is implicated in the genesis and non-Mendelian propagation of new genetic information besides that contained in chromosomes.'

5

Blind Chance and Natural Selection

*When these deviations only affect the soma, they give rise to temporary
non-hereditary variations; but when they occur in the germ-plasm,
they are transmitted to the next generation and cause corresponding
hereditary variations in the body.*
(August Weismann, 1892, author of germ-plasm theory,
Keimplasma: Eine Theorie der Vererbung)

Charles Darwin and his Predecessors

Charles Darwin published his ground-breaking work *The Origin of
Species* in 1859. The book was finished in a hurry, but the ideas that he
developed into his theory of evolution by natural selection had slowly
matured through three decades since his famous voyage on *The Beagle*
in 1831–1836 (Figure 5.1).[1]

In South America and its offshore islands, particularly the Galapagos
Islands straddling the equator in the Pacific Ocean, he saw the astonishing variety of living organisms and how they differed from what he
knew in Europe. They differed even between the different islands. The
birds and reptiles on the separate islands were particularly convincing.
They differed in ways that strongly suggested a common origin followed
by adaptation and drift in the environment of each island. The beaks
of birds, for example, changed according to the kind of food available.
The finches were clearly well-adapted to the particular environment of
each island. So were the reptiles. He knew also that plasticity of a similar

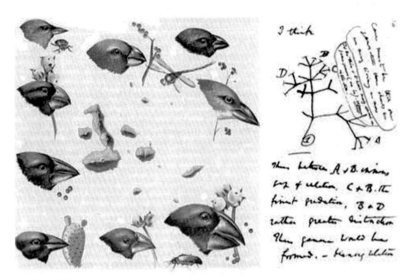

Figure 5.1 Adaptive radiation in the finches of the Galapagos islands, with Darwin's diagram of radiation beginning 'I think…'. He observed similar radiation in the tortoises on the different islands (www.biomedware.com/blog/2012/genetic-gis-a-call-and-a-research-agenda/, right: Charles Darwin's 1837 sketch of an evolutionary tree from his *First Notebook on Transmutation of Species* (1837)). For a colour version of this figure, please see the plate section.

kind, showing variations in the detail within a common overall form, had been exploited by animal and plant breeders for thousands of years. By selecting varieties that emphasised some features like size, colour, facial shapes, slenderness or strength, and many others, they had produced by artificial selection an astonishing variety of dogs, cats, cattle, horses, fish and plants. His insight, developed back in England after the voyage of *The Beagle*, was that selection could have created similar variation naturally without human intervention. And so the idea of *natural* selection was born. All it needed was already existing variations on which the environment could act to make some forms more successful than others. He became convinced that different species must have evolved in this way from earlier forms, stretching back to a possible common ancestor.

However, he did not know what may have caused the variations on which natural selection could act. He did not know that Gregor Mendel was already performing his experiments on plant hybridisation and the laws of transmission of inherited characteristics (traits), which were published in 1866, but not widely known until after Darwin's lifetime. In this respect, Darwin's book was not so much about the *origin* of species as about their selection. Nevertheless, the ideas of natural selection, and the evidence from his journeys and observations, were powerful enough for him to produce his great masterpiece.

The book was finally published in a hurry when another naturalist, Alfred Russel Wallace, was hot on his heels with similar ideas.[2] Wallace was a leading figure in the study of the geographical dispersion of animal species and he had made similar discoveries of great variation in the Malay Archipelago on the other side of the world. This vast group of islands is also tropical. Moreover, it consists of two groups of islands separated by what became called the Wallace line. One group must have been populated by animals from the Australian region, the other by South-East Asian ones, leading to very different origins of the species. This striking fact and the other differences between the islands led Wallace also to the conclusion that natural selection could be the explanation. Justice was served by the Linnean Society of London, where an 1858 paper of Wallace was presented at the same time as an unpublished paper of Charles Darwin that he had written in 1844.[3] That paper is the reason why Darwin was judged to have priority in developing the idea, and Wallace graciously accepted the outcome.[4]

As often happens when a work of many years is completed in a hurry, Darwin had cause to regret the final rush. In particular, he had not had time to write a proper acknowledgement of previous work, even though he knew that such existed. His predecessors even included his grandfather, Erasmus Darwin. So Charles Darwin knew that he was by no means the first to propose a theory of the transformation of species, as evolution was called in his time. Darwin corrected this omission in the third edition of the *Origin*, where he acknowledged no fewer than 30 predecessors. He was generous in his selection. Even Aristotle was listed. In the fourth edition the list had grown to 38.[5] Prominent in this list was Jean-Baptiste Lamarck, of whom Darwin wrote 'this justly celebrated naturalist . . . who upholds the doctrine that all species, including man, are descended from other species'.

He had good reason to praise Lamarck. Half a century earlier in his *Philosophie Zoologique*, published in 1809, exactly 50 years before Darwin's *Origin*, Lamarck had laid out the reasons for transformationism, which he had to defend against severe critics amongst his scientific colleagues, just as Darwin would have to do. Darwin was greatly helped in this task by Thomas Henry Huxley, 'Darwin's bulldog', who acted as the public face of Darwinism, while Darwin himself worked quietly away at his country home. Lamarck had to fight his battles in the intellectually challenging Parisian culture, largely alone against powerful opponents like Georges Cuvier, who wrote an obituary oration that systematically trashed Lamarck's reputation. That oration was read at Lamarck's pauper burial and it was to reverberate like a death knell across time, as we will see in Chapter 6. It was written from a highly biased perspective. Cuvier believed in a form of creationism which proposed that new species were separately created following global catastrophes, and he was strongly opposed to Lamarck's ideas.

Darwin on Lamarckism

But the championing of transformationism was not the only reason Darwin had for praising Lamarck. Lamarck is also known for having embraced the idea of the inheritance of acquired characteristics and developed it into a theory of transformation of species. Amongst those who accepted transformationism, this was a common view. Use and disuse of organs of the body were thought to be the cause of inherited changes. Fish living in unlit caves had become blind through not using their eyes, giraffes had grown long necks through reaching up to higher trees for food. This idea became known as Lamarckism because it was the mechanism of variation assumed, but not invented, by Lamarck. I think Lamarck would be surprised to find it called Lamarckism, but we are now stuck with that word so I will use it freely in this book.[6]

Many people today believe that Darwin was opposed to Lamarck's idea. That is not correct. On the contrary, there are 12 places in the *Origin* where this process is referred to. In fact, he went even further than Lamarck. He proposed a biological mechanism for how it may happen. This is not difficult to envisage in unicellular organisms, where the cell that forms the whole organism divides to reproduce. Changes in the structure of the cell can therefore be inherited even if the genetic material is not changed. The first experimental example of this process was discovered by Tracy Sonneborn, working in the USA in the 1960s, who showed that changing the orientation of the swimming cilia in a part of the animal by surgery produced an inherited characteristic, transmitted down at least two subsequent generations.

This kind of inheritance is also easier to imagine in plants, many of which can reproduce from any part of the plant. Indeed, I think that such inheritance must occur in such organisms.[7] To exclude the inheritance of acquired characteristics a priori must be misguided since much more is inherited than the genetic database that we now know is DNA.

That is necessarily true for all single-cell organisms. They form the vast majority of life on Earth. Remember, too, that for most of the period of evolution those were the only organisms that existed. Before the evolution of multicellular organisms with a separate germ line of cells distinct from the somatic cells, there wasn't a problem. It is a little-appreciated fact that the Modern Synthesis which will be analysed later in this chapter could only ever have been properly applied to a relatively small period of the evolutionary process, perhaps no more than 20%.

The problem that Darwin saw is that it is not obvious how such a process could happen in multicellular organisms possessing a separate germ line. If changes in the soma occur as a consequence of adaptation to the environment, there is no reason why this should change the germ line cells unless the adaptation is somehow transmitted to them. How, for example, could an adaptation in a sense organ like the eye, forming part of the soma, have an effect on sperm or egg cells far away in the reproductive organs containing the germ cells? Something would have to transmit the information to the germ cells so that it could be inherited by the next generation. To deal with this problem Darwin invented the idea of gemmules, little particles that he supposed to travel through the blood to carry the relevant influences.[8] As we will see in Chapter 8, this idea is not very far away from the evidence on what we know today about possible mechanisms for this form of inheritance.[9]

Perhaps one reason why Darwin continued to include a Lamarckian process was that, deep down, he must have been unhappy that he did not know the mechanism of variation and that selection alone could not provide that. Darwin was a worrier, never fully happy with his Ideas.[10]

For these reasons, amongst others, Darwin's view of evolution was a fairly nuanced one with multiple mechanisms, prominent amongst which was natural selection. One reason for that caution is that people at that time knew nothing about genetics. He wrote in the first edition of the Origin: 'I am convinced that natural selection has been the main, *but not the exclusive* means of modification',[11] a statement he reiterated with increased force in the sixth edition in 1872. The caution in avoiding the claim that natural selection was the *only* process involved is characteristic of Darwin, who often expressed his own doubts and uncertainties. Sadly, this cautious humility has not been shared by many who claim to follow in his footsteps. It is an important part of the reasons for which Darwin was not a Neo-Darwinist.

The Rise of Neo-Darwinism Leading to the Modern Synthesis

As we will see in this chapter, Neo-Darwinism is significantly different from Darwin's more nuanced position and therefore needs to be clearly distinguished from Darwinism. Many Neo-Darwinists today fail to acknowledge this and readily imply that they are Darwinists. Neo-Darwinism was first introduced as a term by the physiologist George Romanes in a letter to *Nature* in 1895,[12] but the roots go back to August Weismann in around 1870. Neo-Darwinism led to a narrowing down of the mechanisms supposed to have been involved in the origin of different species. The motivation for the development of Neo-Darwinism was a move to exclude Lamarckism. Despite Darwin's acceptance of the idea, the inheritance of acquired characteristics was deemed impossible. Natural selection working on chance variations in genetic material was thought to be entirely sufficient to explain all evolutionary change.

To explain how this narrow interpretation of Darwin's ideas came about, we need to turn to early experiments to test Lamarckism.

As we have seen, Darwin's work was done with no knowledge of genetics. Mendel's work on the genetics of hybridisation in peas was not rediscovered until the late nineteenth century, when the first experiments to test the inheritance of acquired characteristics were also done.

August Weismann was the first to do this in 1890. He decided to do so not by exposing animals, and particularly their embryos, to different environments, but rather by treating them surgically. This fact is crucial. The experiments consisted in amputating the tails of mice and then observing whether this had any effect on the progeny. The answer was a clear 'no'. Since this work forms a foundation stone of Neo-Darwinism, it is important to ask whether it really answers the relevant question. We can then return to the development of Neo-Darwinism.

Even on the older versions of Lamarckism, as expressed by Lamarck himself, this is a curiously inappropriate way of testing the idea. The idea is that Lamarckian inheritance may occur in a *functional* interaction between the organisms *and their environment*, through use and disuse of the organism's structures and functions, not whether the *non-functional results of surgery* can be inherited. Darwin must have known already from the work of animal breeders that such inheritance did not occur. Tail amputation in dogs for aesthetic reasons does not result in stunted tails in the offspring, no matter how many generations are bred from the animals. To put the question in a more modern form, it is whether the germ line is or is not isolated from environmental influences. The relevant way to do a tail-cutting experiment or any other experiment to answer that question would be to change the environment in a way that makes tail-lessness a functional advantage. Quite apart from the obvious question of why a surgical change should be inherited, even a standard Lamarckian would notice that the environment, apart from the surgery, is not different. Furthermore, even if there were environments that would favour tail-lessness, the experiment would not test for that. The work of Conrad Waddington in Chapter 8 will show the more successful way forward for such experiments.

Nevertheless, the experiment convinced Weismann and others that Lamarckism is impossible. The Weismann Barrier, i.e. the isolation of the germ cells from the soma, then became a cornerstone of the development of Neo-Darwinism into the Modern Synthesis.

At its roots, therefore, Neo-Darwinism, in relying on the Weismann Barrier so completely as a kind of dogma, made a serious mistake that could have been avoided.[13] As we will see, as Neo-Darwinism developed into the Modern Synthesis, it also made a further mistake in embracing the molecular biological equivalent of the Weismann Barrier, which is the Central Dogma of Molecular Biology.[14] Unnecessary dogmatism has unfortunately haunted evolutionary biology.[15]

Weismann is also credited with the idea, which he developed in his *Essays upon Heredity* as early as 1889, that changes in the germ line cells were largely random, which also became a kind of dogma. He was therefore responsible for the two main assumptions of the Modern Synthesis and it is not surprising that he is often judged to be the most important evolutionary biology thinker forming the link between Darwin and the formulation of the Modern Synthesis in the 1930s and 1940s. Ernst Mayr, author of the magisterial 1982 book *The Growth of Biological Thought*, described him as 'one of the great biologists of all time'.

It is curious, to say the least, that such a distinguished and widely praised scientist should have put up such an inappropriate straw man to knock down, and even more curious that so many other eminent scientists should have accepted it as the basis of a major evolutionary theory. No major publishable science today could be based on such flimsy and inappropriate evidence.[16] Why did Weismann and his successors do that?

In Weismann's case, part of the answer lies in the little-known fact that he fully acknowledged that his experiments disproved only the inheritance of surgical mutilation.[17] He performed these experiments because he was aware of 'Lamarckian' claims that such mutilations could be inherited. The straw man had already been set up by over-imaginative Lamarckians, who had claimed, for example, that repeated circumcision in the generations of populations that practise this particular surgery could lead to babies born without foreskins. He well knew the limitations of his experiments and that their crudeness was responding to a similarly crude alternative. I suspect, therefore, that he must have had other reasons for being so sure of the barrier idea. He may already have been convinced of the correctness of his other assumption, the randomness of variations. And he delivered an important lecture in 1883 ('On inheritance') in which he rejected the inheritance of acquired characteristics and proposed alternative explanations for the use and disuse examples Darwin gave in the *Origin*. But that still leaves the question of why so many other leading scientists followed in his footsteps and enshrined the Weismann Barrier into the Modern Synthesis. The mistake is not a subtle one. Mutilations simply do not test a functional interaction between an organism and the environment. Nor does showing that there are alternative explanations for Darwin's examples prove that they are the correct explanations. So far as I am aware, the tail-cutting experiments he performed are the only direct experimental tests.

The answer to that puzzle lies in the names of two great scientists: Gregor Mendel, the originator of modern genetics, and Erwin Schrödinger, the great quantum mechanics pioneer.

Evolution and Genetics

Darwin's work was done with no knowledge of genetics, nor of the remarkable quantitative results being obtained in a European monastery in what today is the Czech Republic, even as the *Origin* was published. Weismann's early experiments were also performed without knowing of Gregor Mendel's work. But soon after that was rediscovered in 1900 many noticed that it married well with Weismann's ideas and those of the developing concept of Neo-Darwinism.

Mendel was interested in the hybridisation of pea plants in the monastery garden. He crossed plants with different characteristics in the peas, pods and flowers. He noticed two facts that have become standard in the teaching of genetics. The first is that, often, characteristics are inherited in an all-or-nothing manner. Peas were either smooth or wrinkled, not usually something between. Other characteristics also showed this behaviour. Yet, a plant showing one characteristic could still give rise to plants with the other characteristic, even if crossed with a plant with the same characteristic. He came to the conclusion that there must be hidden genetic elements and that some were dominant and some recessive. In a plant with genes for both, only the dominant characteristic would be expressed. This is a common feature of inheritance in many species. Two humans with the same hair colour, say black, can have progeny with ginger hair. The dominant–recessive idea can easily explain this kind of result. Both parents must have had recessive ginger genes. Since sexual reproduction transmits two copies of the genetic material to the progeny, the recessive characteristic will only be manifest if both contain the gene for this characteristic. There is no blending. The same was thought to be true of eye colour in mammals. In humans, we find either blue or brown, not brownish blue or bluish brown.[18]

This discovery was the basis of the development of an atomistic view of genetics. Genes can act seemingly as single 'atomic' elements to produce the phenotype characteristic, and they do so in an all-or-nothing manner. This leaves no room for Lamarckian blending. Although Weismann was initially sceptical of Mendel's work, his followers readily saw it as providing the necessary boost to otherwise flimsy evidence for the impossibility of Lamarckian inheritance. This partly explains the reason why the flimsy surgical evidence was deemed sufficient, though not on its own.

The next step in Mendel's work laid the foundations for quantitative genetics, and for all the wonderful advances that were to come during and following the formulation of the Modern Synthesis in what became called population genetics. Evolutionary biology could at last become mathematical.

The idea is simple to explain. Mendel saw that if it required parents with both dominant and recessive copies of the gene involved, and if the inheritance is all-or-nothing, then it is easy to predict the ratio in which the progeny should show the recessive character. Let's call the forms D and R. Today we call these alleles. They occupy the same position in the genome and substitute for each other in the sequence. Mendel could not have known that, of course, but he nevertheless saw that the process was 'atomistic' and 'quantal' in expression. If the D and R forms in parents that have both are mixed randomly during sexual reproduction, then the chances of the following forms in the offspring would all be equally likely to occur: DD, DR, RD and RR. But only RR would allow R to be expressed. So, it would appear just 25% of the time. Mendel's second great discovery was that this is what happens.[19]

To some modern eyes, steeped in the maths of this kind of probability calculation and in the uncertainties that must be present in any experimental science using probabilities, Mendel's results have seemed just a little bit too good to be entirely true. Perhaps he had reasons for excluding some data. All science does that to some extent. There are nearly always rogue results. The difficulty lies in establishing really sound criteria for excluding any data. Nevertheless, no one doubts that Mendel's insight, and the broad sweep of his results, were correct.[20]

Notice also the assumption of randomness here. If genetic mixing in sexual reproduction is a random process, then perhaps all variations are random. It is easy to see why Mendel's work bolstered Weismann's ideas. And so, the incorporation of Mendelian mathematics into Neo-Darwinism led to the full formulation of the Modern Synthesis during the first half of the twentieth century, leading up to the publication of Julian Huxley's book, *The Modern Synthesis*, in 1942.

The Modern Synthesis

Mendel's genetic experiments were unknown when the term Neo-Darwinism was coined at the end of the nineteenth century. One of the reasons why Neo-Darwinism developed into what is called the Modern Synthesis is the integration of Mendelian genetics into the Neo-Darwinist framework.

Many biological scientists today who claim to adhere to the Modern Synthesis and therefore describe themselves as Neo-Darwinists actually accept some or even all of the exceptions and breaks with Neo-Darwinism that I will describe in Chapters 7 and 8. Neo-Darwinism (recall that I am using the terms interchangeably because that is what many Neo-Darwinists do today) was such a successful theory in terms of the areas of science that it spawned, and it chimed so well with what were thought to be secure discoveries in molecular biology, that it is understandable that many people would prefer to take the view that it can readily accommodate exceptions and new mechanisms. With a little bit of underplaying on how important the exceptions might be, and how they might be the exceptions that 'prove the rule', this is a tactic often adopted during paradigmatic shifts in understanding.[21]

This book argues that it is not sufficient to accommodate the exceptions as minor in this way. To make it clear why I take that view it is helpful to summarise what I see as the original central ideas of Neo-Darwinism before analysing the philosophical ideas that underpin it. They correspond very well indeed with the major popularisations of the theory, including Richard Dawkins' 1976 book *The Selfish Gene*, which has not only popularised the theory, but also spawned many applications, implications and consequences in fields as diverse as philosophy, economics, theology, sociology, politics and literature, as well as within the standard biological and physical sciences. It is hard to think of a scientific popularisation that has had as wide and deep an impact as this, which is a tribute to the highly successful style in which Dawkins writes his many books.

The Original Version

The main ideas in the original formulation of Neo-Darwinism are:

1. All changes in the genetic material are random, as Weismann first proposed. It is important to note that what is meant is that genetic change is random with respect to function. This is the 'blind chance' part of the theory with no room for teleology.

2. The germ line cells are completely isolated from the rest of the organism. Dawkins encapsulated this view in *The Selfish Gene*: 'Sealed off from the outside world.'

3. Development is simply an unfolding of what is in the DNA 'blueprint'. The great twentieth-century expounder of Neo-Darwinism Ernst Mayr put it very succinctly in 1982: 'All of the directions, controls and constraints of the developmental machinery are laid down in the blueprint of the DNA genotype as instructions or potentialities.'

4. The random variations may produce organisms that are less or more fit in the environment they find themselves in.

5. The more fit organisms produce more offspring that are fertile than the less fit ones.

6. This will therefore lead to the genes in the fitter individuals gradually coming to dominate the gene pool of the population. This is natural selection, Darwin's central idea.

There are also some negative statements to add to the list. Most important is the exclusion of Lamarckian forms of inheritance. Some modern defenders of Neo-Darwinism claim to accommodate the Lamarckian forms that have now been discovered, but this seems to me to be contrary to common sense.[22] The central assumption of Neo-Darwinism is that the inheritance of acquired characteristics is impossible. But if inheritable variations are not always random with respect to physiological function, then it seems to me to be more honest to say so.

A similarly strong negative statement reinforces the separation between evolution and development. To quote Mayr again: 'The clarification of the biochemical mechanisms by which the genetic program is translated into the phenotype tells us absolutely nothing about the steps by which natural selection has built up the particular genetic program.'

Only from this dogmatic gene-centred theory is it possible to view Neo-Darwinism as very economical as a theory. With essentially two assumptions, random variation and natural selection, it seemed that everything could be explained. As Dawkins said in a debate in 2009 with one of his antagonists, the great pioneer of the process called symbiogenesis, Lynn Margulis:

> It's highly plausible, it's economical, it's parsimonious, why on earth would you want to drag in symbiogenesis when it's such an unparsimonious, uneconomical [theory]?

I will leave Margulis' reply to later in this chapter. All we need to note at the moment is that being parsimonious is not a guarantee of truth. Parsimony is a relativistic concept, dependent on the scale we are addressing. Moreover, as we saw in Chapter 1, quantum mechanics and relativity theory greatly complicated the simple nineteenth-century view of the physics of the universe. Being parsimonious didn't save it from being falsified, however strongly and convincingly Laplace and others had presented it. Something similar is now happening to biological science. The full details of that process will appear in Chapters 7 and 8.

Extensions of the Modern Synthesis

It is important to acknowledge that evolutionary biologists have added new processes to this list since the Modern Synthesis was formulated around 75 years ago. The discoveries that the list is incomplete occurred in various stages over the subsequent years, and these led to various extensions.

The first extensions were based on the realisation that evolutionary change could occur without any mutations, and even if there was no selection pressure. Purely random variations in selection of existing gene variants (alleles) can give rise to a process that is called random genetic drift.[23]

Genetic drift was extended around 1968 when the Japanese scientist Motoo Kimura developed the idea into the neutral theory of molecular evolution.[24] This theory assumed that the great majority of the differences in genome sequences were neutral with respect to function and so would not affect the fitness of the individual and would not therefore be selected for or against. Most evolutionary change was therefore attributed to genetic drift. Selection was allocated the role of removing harmful mutations.

Neither of these developments altered the basic assumption that genetic variation was random. They were therefore relatively easy to incorporate into the Modern Synthesis.

A more recent set of extensions has been to incorporate developmental biology back into evolutionary theory, leading to what is widely called evolutionary development biology (evo-devo for short). This theory is more important to the theme of this book since it proposes that evolutionary change has occurred as a result of changes in genetic expression patterns or through genes acquiring different roles in new functions. In the form favoured by the American scientist Mary Jane West-Eberhard, it also led to the idea that phenotype change can occur before genotype change.[25] The idea that genes are followers rather than leaders in evolution will be explored further in Chapter 8, where one of the questions will be whether the Modern Synthesis can continue to be extended or whether it is time for replacement.

Schrödinger and *What is Life?*

Erwin Schrödinger is best known for his work on quantum mechanics, where he famously formulated the wave equation. He also worked on relativity theory and on ways of unifying quantum mechanical and relativity views of the universe. He gave a series of lectures at the Dublin Institute of Advanced Studies in 1942, the same year in which Julian Huxley's book was published. Those lectures were published in 1943 with the ambitious title *What is Life?*. It was a seminal adjunct to Huxley's book in two important respects. He made two predictions, one of which was fulfilled spectacularly; the other is necessarily incorrect.

At this point in the story it is important to note that while Mendel's work showed that there must be genetic material somewhere within the organism and that inheritance could be discrete, no one knew where the genetic material would be found. When proteins were found to be amino acid polymers, they might have been a suitable candidate. Amino acids can be organised in billions of different sequences, so that would have been quite possible from an information perspective. Locating it in the other main biological polymer, DNA, was yet to come. Yet Schrödinger made the completely correct prediction that, wherever it was found, it would turn out to be what he called an aperiodic crystal. A crystal can be defined as molecules arranged in regular orders. In a slightly stretched sense, a polymer is that, even if it is more like a string than what we normally think of as a crystal. 'Aperiodic crystal' is also a very good description since, although there are repeats in the DNA sequences, in fact many of them, they are not regularly repeated throughout the genome. They are usually short sections in an otherwise information-dense aperiodic sequence. Schrödinger realised that if the genetic molecules were to contain sufficient information to enable an organism to develop from a single cell, then the structural sequences of the 'crystal' would need to be

irregular. A regular repeating structure would contain very little information over and above the first instance of it. This is true of all sequences.
The words of this book, for example, contain (I hope!) a lot of useful
information for you as the reader. But you would be very disappointed
to open the book to find that it consisted of the first sentence endlessly
repeated.

Now we come to Schrödinger's second conclusion. If the genetic
material is an information-dense sequence, how is it read to enable
the characteristics of an organism to be transmitted from one generation to another? A one-dimensional sequence cannot simply map a
three-dimensional structure. It is not a miniature organism in the way
in which some nineteenth-century microscopists imagined when they
looked at sperm and egg cells. Could that three-dimensional template
come from somewhere else, perhaps in the three-dimensional structure
of the cell itself? Whichever way that is done, Schrödinger reasoned that
the sequence must be read in a determinate manner if it was faithfully
to transmit information. Stochasticity in a communication line is intol-
erable. We call it noise, and if loud enough it can drown out the signal.
From this he concluded that there must be an absolutely fundamental
difference between physics and biology.

Physics can be characterised as order from disorder. At the micro
level, there is the essential stochasticity of quantum mechanics. Even
if, one day, an alternative view of 'reality' is produced, as people like
Albert Einstein and David Bohm believed, we can't escape the fact that
the equations of quantum mechanics are precisely predictive as proba-
bilistic descriptions. Any underlying determinism would have to reproduce this. That is not difficult to imagine, since we already have an
example of stochasticity at the molecular level that was discovered well
before quantum mechanics. In 1827 Robert Brown observed that fine
particles derived from pollen grains showed stochastic movement in
water observed under the microscope. We call it Brownian motion and
it was shown by Einstein in 1905 to arise from the random bombard-
ment of the particles by the random motion of water molecules, the first
demonstration of the existence of individual molecules with separate
motions.

Yet, the equations of thermodynamics, which describe large numbers of particles to generate the gas laws, are determinate. The answer to
this apparent paradox is that, if motion at the particle level is genuinely

random, then large numbers of particles will cancel their individual movements out to produce a constant pressure when hitting an object, like the wall of a pressure vessel. Order at large scales therefore arises from disorder at lower scales.

But this interpretation is inconsistent with a Schrödinger view of biology in which the genetic material at the molecular level is supposed to be read in a determinate manner, rather as an X-ray beam can generate an accurate and determinate 'picture' of a crystal by the diffraction of the rays by the regular structure of the crystal. Biology, he reasoned, was therefore the generation of order at large scale from order at the micro scale.[26]

This can't be correct, as we will see later in the story. Moreover, it is strange given his background and great contributions to relativity theory that Schrödinger seems to have seen no reason to view biology as the next area of application of the general principle of relativity. His view of biology as generating order from order across the scales must have shielded him from such insights. It also led to the incorrect 'read-only' view of DNA.

Neo-Darwinism and the Central Dogma

The scientists who developed the Modern Synthesis were brilliant and they were amongst the most influential scientists of the twentieth century. They formulated the best hypothesis they could that would combine the observations and insights of Darwin and Wallace on the role of natural selection with the discoveries of Mendel and the idea of the Weismann Barrier on genetics. As a hypothesis, it was very successful. The study of the genetics of populations was transformed and became a rigorously mathematical discipline.

The Modern Synthesis also fitted extremely well with the early discoveries of molecular biology.[27] Schrödinger's aperiodic crystal has become the genome. Following the determination of the structure of DNA by James Watson, Francis Crick, Maurice Wilkins and Rosalind Franklin in 1953, it became clear that not only was the molecule a one-dimensional string coiled into a helical structure, it was also impossible for there to be a one-to-one mapping between nucleotides of DNA and the amino acids of proteins. There are only four nucleotides used in DNA; 20 amino acids

nonpolar	polar	basic	acidic	(stop codon)

RNA templates for amino acids

1st base	2nd base				3rd base
	U	**C**	**A**	**G**	
U	UUU (Phe/F) Phenylalanine UUC UUA (Leu/L) Leucine UUG	UCU (Ser/S) Serine UCC UCA UCG	UAU (Tyr/Y) Tyrosine UAC UAA Stop (Ochre) UAG Stop (Amber)	UGU (Cys/C) Cysteine UGC UGA Stop (Opal) UGG (Trp/W) Tryptophan	U C A G
C	CUU (Leu/L) Leucine CUC CUA CUG	CCU (Pro/P) Proline CCC CCA CCG	CAU (His/H) Histidine CAC CAA (Gln/Q) Glutamine CAG	CGU (Arg/R) Arginine CGC CGA CGG	U C A G
A	AUU (Ile/I) Isoleucine AUC AUA AUG(A) (Met/M) Methionine	ACU (Thr/T) Threonine ACC ACA ACG	AAU (Asn/N) Asparagine AAC AAA (Lys/K) Lysine AAG	AGU (Ser/S) Serine AGC AGA (Arg/R) Arginine AGG	U C A G
G	GUU (Val/V) Valine GUC GUA GUG	GCU (Ala/A) Alanine GCC GCA GCG	GAU (Asp/D) Aspartic acid GAC GAA (Glu/E) Glutamic acid GAG	GGU (Gly/G) Glycine GGC GGA GGG	U C A G

Figure 5.2 The triplet code. There are only 20 amino acids normally used in proteins. There are 64 different triplet sequences using four bases. Many of the triplet sequences therefore code for the same amino acid. Three triplet sequences, UAA, UAG and UGA, are used to indicate stop points (adapted from Wikipedia, http://en.wikipedia.org/wiki/Genetic_code, Mikael Häggström (Public Domain licence)).

are used to make proteins. The minimum sequence to specify an amino acid would therefore be three, since two would only allow 16 to be coded for. This was subsequently shown to be the case, and so the triplet code was discovered (Figure 5.2).

Molecular biology was also successful in working out the way in which the sequences are read. But the reading is one-way only. Nucleic acid sequences code for amino acid sequences but not the other way round. This is the basis of the formulation of the Central Dogma of molecular biology, which states exactly that.

So, with all this success to its name, what went wrong with Neo-Darwinism? In a single word: hubris.

What went wrong was that the Modern Synthesis became hardened into dogmatism.[28] Starting from the theory that this is the way in which evolution *could* have happened, it became transformed into the conviction that this was the only way in which evolution *must* have happened. I referred above to the 2009 debate between Richard Dawkins and Lynn Margulis. Margulis was the champion of the role of symbiogenesis in the formation of eukaryotes via the fusion of bacteria with early cells. I will now give a slightly longer section of the transcript of that part of the debate:

DAWKINS: It [Neo-Darwinism] is highly plausible, it's economical, it's parsimonious, why on earth would you want to drag in symbiogenesis when it's such an unparsimonious, uneconomical [theory]?
MARGULIS: Because it's there.

That's it in a nutshell. What is there, what exists, is the starting point of all science.

Through many such starting points we have learnt that Nature often resists parsimonious explanations at low levels. As we saw in the case of cardiac and circadian rhythms in Chapter 3, multiple processes are the rule rather than the exception. It would also have been incorrect to have insisted that the first 'clock' gene to be discovered was necessarily the only one to be involved in circadian rhythm. It is similarly incorrect to insist that the Neo-Darwinist mechanism is the only one required to explain evolution. As we have seen earlier in this chapter, Darwin himself did not think that natural selection was the only mechanism.

The Language of Neo-Darwinism[29]

The problems, though, run even deeper than the new (and older) experimental evidence that the theory is incomplete in describing the known facts. Many of the problems with the standard theory of evolution in accommodating the new experimental findings have their origin in Neo-Darwinist metaphors and other forms of representation rather than in experimental biology itself. These colourful metaphors have been responsible for, and express, the way in which twentieth-century biology has most frequently been interpreted and presented to the public. In addition, therefore, to the need to accommodate unanticipated

experimental findings, we need to review the way in which we interpret and communicate experimental biology. The language of Neo-
Darwinism and twentieth-century biology reflects highly reductionist
philosophical and scientific viewpoints, the concepts of which are not
required by the scientific discoveries themselves. In fact, it can be shown
that, in the case of some of the central concepts of Neo-Darwinism, such
as 'selfish genes' or 'genetic programme', no biological experiment could
possibly distinguish even between completely opposite conceptual interpretations of the same experimental findings.[30] The concepts therefore
form a biased interpretive veneer that can hide those discoveries in a web
of interpretation.

I refer to a web of interpretation since it is the whole conceptual
scheme of Neo-Darwinism that creates the difficulty. Each concept and
metaphor reinforces the overall mind-set until it is almost impossible
to stand outside it and to appreciate how beguiling it is. Since Neo-
Darwinism has dominated biological science for over half a century, its
viewpoint is now so embedded in the scientific literature, including standard school and university textbooks, that many biological scientists may
themselves not recognise its conceptual nature, let alone question inco-
herencies or identify flaws. Many see it as merely a description of what
experimental work has shown: the idea in a nutshell is that genes code
for proteins that form organisms via a genetic programme inherited from
preceding generations and which defines and determines the organism
and its future offspring. What is wrong with that? The remainder of this
chapter explains what is wrong or misleading and, above all, it shows that
the conceptual scheme is neither required by nor any longer productive
for, the experimental science itself. Nor is it consistent with the principle
of relativity applied to multi-scale biology.

I will analyse the main concepts and the associated metaphors indi-
vidually, and then show how they link together to form the complete
narrative. We can then move on in Chapters 7 and 8 to ask what
would be an alternative approach better fitted to what we now know
experimentally and to a new more integrated systems view of the processes of living organisms and of their evolution. The terms that require
analysis are 'gene', 'selfish', 'code', 'programme', 'blueprint' and 'book of
life'. We also need to examine secondary concepts like 'replicator' and
'vehicle'.

'Gene'

As we have seen, Neo-Darwinism is a gene-centred theory of evolution. Yet its central notion, the 'gene', is an unstable concept. Surprising as it may seem, there is no single agreed definition of 'gene'.[31] Even more seriously, the different definitions have incompatible consequences for the theory.[32]

Johannsen's original definition, GeneJ – the necessary cause of a trait. The word 'gene' was introduced by Johannsen in 1909. But the concept had already existed since Mendel's experiments on plant hybrids, published in 1866, and was based on what one of the great thinkers of Neo-Darwinism, Ernst Mayr, described in 1982 as 'the silent assumption [that] was made almost universally that there is a 1:1 relation between genetic factor (gene) and character'. Of course, no one now thinks that there is a simple 1:1 relation, but the language of direct causation has been retained.

I will call this definition of a 'gene' geneJ to signify Johannsen's (but essentially also Mendel's) meaning. Since then, the concept of a gene has changed fundamentally. GeneJ referred to the cause of a specific inheritable phenotype characteristic (trait), such as eye/hair/skin colour, body shape and weight, number of legs/arms/wings, to which we could perhaps add more complex traits such as intelligence, personality and sexuality.

The modern molecular biological definition of a gene, GeneM – a sequence of DNA that is transcribed to produce a functional product. The molecular biological definition of a gene is very different. Following the discovery that DNA forms templates for proteins, the definition shifted to locatable DNA sequences with identifiable beginnings and endings. Complexity was added through the discovery of regulatory elements (essentially switches), but the basic cause of phenotype characteristics was still thought to be the DNA sequence, since that forms the template to determine which protein is made, which in turn interacts with the rest of the organism to produce the phenotype. I will call this definition of a 'gene' geneM (see Figure 5.3).

The important point here is that geneJ and geneM are not the same thing and their causal relationships to the phenotype are fundamentally different.[34] They could only be the same thing if all phenotype

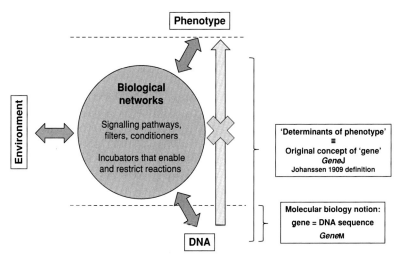

Figure 5.3 Relations between genes, environment and phenotype characters according to current physiological and biochemical understanding. This diagram represents the interaction between DNA sequences, environment and phenotype as occurring through biological networks. The causation occurs in both directions between all three influences on the networks. This view is very different from the idea that genes 'cause' the phenotype (right hand arrow). This diagram also helps to explain the difference between the original concept of a gene as the cause of a particular phenotype (geneJ) and the modern definition as a DNA sequence (geneM).[33]

characteristics were attributable entirely to DNA sequences. As we have seen in Chapter 4, this is not the case. DNA does not act outside the context of a complete cell. The complete cell is also necessary.

The causal relationships are also completely different. According to the original view, genesJ were *necessarily* the cause of inheritable phenotypes since that is how they were defined: as whatever in the organism is the cause of that phenotype. Johannsen even left the question of what that might be vague: 'the gene was something very uncertain, "ein Etwas" ("a something" in German), with no connection to the chromosomes'. Dawkins, in 1982, also used this 'catch-all' definition as 'an inheritable unit'.[35] It would not matter whether that was DNA or something else or any combination of factors. No experiment could disprove a 'catch-all' concept since any new discovery as a cause of the phenotype to be included would be welcomed as a geneJ. The idea becomes completely unfalsifiable. It cannot be tested.

By contrast, defining genes as sequences of DNA makes the question of causation become an empirical investigation. To appreciate the difference, consider Mendel's experiments showing specific phenotypes, such as smooth or wrinkled surfaces of peas. GeneJ was whatever in the plant caused the peas to be smooth or wrinkled. It would not make sense to ask whether geneJ was the cause. That is how it was defined. It simply *is* everything that determines the inherited phenotype, i.e. the trait. Of course, different questions of an empirical nature could be asked about genesJ, such as whether they follow Mendel's laws. Some do; some don't. By contrast it makes perfect sense to ask whether a specific DNA sequence, geneM, is involved in determining the phenotype. That question is open to experimental investigation. GeneJ could only be the same as geneM if DNA alone determined the phenotype.

This difference between geneJ (which refers to indeterminate entities that are *necessarily* the cause) and geneM (whose causation is open to *experimentation*) is central and I will use it several times in this book. The difference is in fact large since most changes in DNA do not necessarily cause a change in phenotype. Organisms are very good at buffering themselves against genomic change – 80% of gene (meaning DNA) knockouts in yeast, for example, are normally silent, while critical biological oscillators like the cardiac pacemaker or circadian rhythm are buffered against genomic change through extensive back-up mechanisms, as we saw in Chapter 3.

The original concept of a gene has therefore been adopted, but then significantly changed by molecular biology. This led to a great clarification of molecular mechanisms, surely one of the greatest triumphs of twentieth-century biology, and widely acknowledged as such. But the more philosophical consequences of this change for higher-level biology are profound and they are much less widely understood. Figure 5.3 illustrates the difference.

Some biological scientists have even given up using the word 'gene', except in inverted commas. The popular idea that a 'gene' is a concrete reality is no longer useful. That change is best illustrated by asking why the precise definition matters.

Why does the definition of a gene matter? A central feature of gene-centred evolutionary theory is that genes are not only the determinants of the organism and the object of selection, they are also isolated from influences of the environment. Genes are seen as unique in being transmitted as 'immortal' elements from one generation to the next via the process of replication. As the replicator they are distinct from the rest of the organism which, from the genes-eye view, is merely the carrier, the 'vehicle', by which they are transmitted.

The reason that the original and the molecular biological definitions of a gene have incompatible consequences for Neo-Darwinism is that only the molecular biological definition, geneM, could be compatible with such a strict separation between the 'replicator' and the 'vehicle'. As illustrated in Figure 5.3, a definition in terms of inheritable phenotypic characteristics (i.e. geneJ) necessarily includes much more than the DNA, so that the distinction between replicator and vehicle is no longer valid. Note also that the change in definition of a gene that I am referring to here is more fundamental than some other changes that are required by recent findings in genomics, such as the 80% of 'non-coding' DNA that is now known to be transcribed and which also might be included in the molecular biological definition. Those findings raise an empirical question: are those transcriptions as RNAs functional? That would extend geneM to include these additional functional sequences. The difference I refer to, by contrast, is a conceptual one. The difference between geneJ and geneM would still be fundamental since it is the difference between necessary and empirically testable causality, not just an extension of the definition of geneM.

This fundamental difference does not seem to have been understood by those who popularised Neo-Darwinism. Yet it is critical to the concepts of 'replicator' and 'vehicle'.

'Selfish Gene'[36]

It is also critical to the colourful metaphor of the 'selfish gene'. On selfish gene theory, genes as DNA sequences are rather like viruses exploiting the carrier organism for their own selfish aims. The metaphor doesn't work for the original definition of a gene since, as we have seen, that included non-DNA parts of the organism. If there is no clear boundary between the gene and the phenotype by which it is expressed, then what is selfishly exploiting what? GeneJ is necessarily defined in terms of the phenotype that is being exploited.

So, can the metaphor work for the molecular definition of a gene? The best way to answer this question is to apply the acid test of any scientific theory, which is to compare it with an opposite theory and ask whether there is an experiment that could be performed to distinguish which is correct. There are various possible opposites that can be used to carry out this test. For example, we could follow Maynard Smith and Eors Szathmary, who wrote in 1999 that 'they [genes] are all in the same boat', which portrays genes as elements that have been imprisoned in the boat and must co-operate to survive.

What we find when we carry out this test is that there is no biological experiment that could distinguish between the selfish gene theory and its opposites, such as 'imprisoned' or 'co-operative' genes. This point was implicitly conceded long ago by Richard Dawkins in his 1982 book *The Extended Phenotype*, where he wrote 'I doubt that there is any experiment that could prove my claim.'[37]

Genes defined as DNA sequences are rather like the symbols of a written language which in themselves have no meaning other than that which is given to them by the users of the language. That is why we are completely baffled by unfamiliar scripts. Until we learn the meanings that are attached to them by the language users, they are mere ciphers. Knowing the letters of the English alphabet does not in itself guarantee that one can understand the works of Shakespeare, or even the talk of a young child. Moreover, the same sequence can have totally different meanings in different contexts within the same language (compare, for example, the English word 'just' in 'just a minute' and 'a just war') and even more so in different languages (compare 'but' in English, translated as 'mais' in French, with 'but' in French, translated as 'goal' in English). DNA sequences are precisely like this. Their meanings in terms of phenotype expression have completely changed during the course of evolution, and they may also change during the life history of a single organism. Many DNA sequences that are necessary for us to have arms and legs first arose in organisms that have neither. The 'language of the genes' has itself evolved. And the key to understanding this 'language' is not at the level of DNA. It is at the level of the phenotype. It is the phenotype that can give meaning (function) to DNA sequences.

Could this problem be avoided by attaching a meaning to 'selfish' as applied to DNA sequences that is independent of meanings in terms of phenotype? For example, we could say that a DNA sequence is 'selfish' to the extent to which its frequency in subsequent generations is increased. This at least would be an objective definition which could be measured in terms of population genetics. But wait a minute! The whole point of the characterisation of a gene as selfish is precisely that this property leads to its success in reproducing itself. We cannot make the prediction of a theory be the basis of the definition of the central element of the theory. If we do that, the theory is empty from the viewpoint of empirical science.[38]

'Code'

After the discovery of the double helical structure of DNA, it was found that each sequence of three bases in DNA or RNA corresponds to a single amino acid in a protein sequence. As we saw earlier in this chapter (Figure 5.2), these triplet patterns are formed from any combination of the four bases U, C, A and G in RNA and T, C, A and G in DNA. They are often described as the genetic 'code', but it is important to understand that this usage of the word 'code' carries overtones that can be confusing.

A code was originally an intentional encryption used by humans to communicate. The genetic 'code' is not intentional in that sense. The word 'code' has unfortunately reinforced the idea that genes are active and even complete causes, in much the same way as a computer is caused to follow the instructions of a computer program. The more neutral word 'template' would be better. Templates are used only when required (activated); they are not themselves active causes. The active causes lie within the cells themselves since they determine the expression patterns for the different cell types and states. These patterns are communicated to the DNA by transcription factors, by methylation patterns and by binding to the tails of histones, all of which influence the pattern and speed of transcription of different parts of the genome. If the word 'instruction' is useful at all, it is rather that the cell instructs the genome. As Barbara McClintock wrote in 1984 after receiving her Nobel Prize, the genome is an 'organ of the cell', not the other way round.

Realising that DNA is under the control of the system has been reinforced by the discovery that cells use different start, stop and splice sites for producing different messenger RNAs from a single DNA sequence. This enables the same sequence to encode different proteins in different cell types and under different conditions.

Representing the direction of causality in biology the wrong way round is therefore confusing and has far-reaching consequences. The causality is circular, acting both ways: passive causality by DNA sequences acting as otherwise inert templates, and active causality by the functional networks of interactions that determine how the genome is activated.

'Programme'

I was one of the first biological scientists to use the early valve-based machines called computers in 1960. They consisted of vast arrays of thousands of valves (a device functionally equivalent to a transistor in which the electrons moved through a vacuum rather than through a semiconductor), requiring large amounts of space and a lot of electricity. There was no screen, no 'windows', and they were programmed with gibberish code written on long rolls of paper or magnetic tape. Their use was restricted to large-scale mathematical calculations of the kind needed in astronomy, particle physics, theoretical chemistry and (very rarely indeed at that time) biology.

The electromagnetic storage space in the machine was small, so the program had to be entered into the machine by running the tape through a tape reader. It is easy therefore for scientists like me to understand why, when the idea of a 'genetic programme' was introduced by the French Nobel laureates Jacques Monod and Francois Jacob, they referred specifically to the way in which early electronic computers were programmed by paper or magnetic tapes: 'The programme is a model borrowed from electronic computers. It equates the genetic material with the magnetic tape of a computer.' The analogy was that DNA 'programmes' the cell, tissues and organs of the body just as the code in a computer program causally determines what the computer does. In principle, the code is independent of the machine that implements it, in the sense that the code itself is sufficient to specify what will happen when the instructions are satisfied. If the program specifies a mathematical computation, for example, it would contain a specification of the computation to be performed in the form of complete algorithms. The problem is that no complete algorithms can be found in the DNA sequences. What we find is better characterised as a mixture of templates and switches. The 'templates' are the triplet sequences that specify the amino acid sequences or the RNA sequences. The 'switches' are the locations on the DNA or histones where transcription factors, methylation and other controlling processes trigger their effects. As a programme, this is incomplete.

Where, then, does the full algorithmic logic of a programme lie? Where, for example, do we find the equivalent of 'IF–THEN–ELSE' type instructions? The answer is in the cell or organism as a whole, not just in the genome. In his 1970 book, *La Logique du Vivant*, Jacob recognised this when he wrote 'the incessant execution of a programme is inseparable from its realisation', and it was expressed even more clearly by Enrico Coen in his 1999 book *The Art of Genes*: 'Organisms are not simply manufactured according to a set of instructions.' Much earlier, in 1986, Adam Wilkins wrote: 'If the genome were a program . . . it would contain instructions on where the information selection process is to begin and it would specify the rules for the successive selection steps. None of that information is in the genome, at least in any form that can be presently recognised.'[39] Thirty years later, after intensive genome sequence analysis, it still has not been found.

Take as an example circadian rhythm as described in Chapter 3. The simplest version of this process depends on a DNA sequence, *Period*, used as a template for the production of a protein PER whose concentration then builds up in the cytoplasm. It diffuses through the nuclear membrane and, as the nuclear level increases, it inhibits the transcription of *Period*. This is a negative feedback loop of the kind that can be represented as implementing a 'programme' like

IF LEVEL X EXCEEDS Y STOP PRODUCING X,
BUT IF LEVEL X IS SMALLER THAN Y
THEN CONTINUE PRODUCING X.

But it is important to note that the implementation of this 'programme' to produce a 24-hour rhythm depends on rates of protein production by ribosomes, rates of change of concentrations within the cytoplasm, rates of transport across the nuclear membrane and speed of interaction with the gene transcription control site (the switch). All of this is necessary to produce a feedback circuit that depends on much more than the genome. It depends also on the intricate cellular, tissue and organ structures that are not specified by DNA sequences, which replicate themselves via self-templating, and which are also essential to inheritance across cell and organism generations.

This is true of all such 'programmes'. To call them 'genetic programmes' or 'gene networks' is to fuel the misconception that all the active causal determination lies in the one-dimensional DNA sequences. It doesn't. It also lies in the three-dimensional static and dynamic structures of the cells, tissues and organs.

The postulate of a 'genetic programme' led to the idea that an organism is fully defined by its genome, whereas in fact the inheritance of cell structure is equally important. Moreover, this structure is specific to different species. Cross-species clones do not generally work. A rare example is work done in China by Yonghua Sun and his colleagues in 2005 using two different species of fish, where the nucleus of one species was used to replace the nucleus in a fertilised egg cell of the other species. The outcome in the anatomy of the adult that resulted from this cross was determined by the cytoplasmic structures and expression patterns of the egg cells, as well as the transferred DNA (Figure 5.4). The basic features of structural organisation both of cells and of multicellular organisms must have been determined by physical constraints before the relevant genomic information was developed.

As with the word 'code', the purpose of this section is to warn against simplistic interpretations of the implications of the word 'programme'. In the extended uses to which the word has been put in biology, and in modern computing science where the concept of a distributed program is normal, 'programme' can be used in many different ways. The point is that such a 'programme' does not lie in the DNA alone. That is also the reason why the concept of a 'genetic programme' is not testable. By necessarily including non-DNA elements, there is no way of determining whether a 'genetic programme' exists. In the limit, when all the relevant components have been added, the 'programme' is the same as the function it is supposed to be programming. The concept then becomes redundant.

'Blueprint'

'Blueprint' is a variation on the idea of a programme. The word suffers from a similar problem to the concept of a 'programme', which is that

Figure 5.4 Cross-species clone. The nucleus of a common carp, *Cyprinus carpio* (middle), was transferred into the enucleated egg cell of a goldfish, *Carassius auratus* (left). The result is a cross-species clone (right) with a vertebral number closer to that of a goldfish (26–28) than of a carp (33–36) and with a more rounded body than a carp. The bottom illustrations are X-ray images of the animals in the top illustration (figure kindly provided by Professor Yonghua Sun from the work of Sun *et al.* (2005)).[40] For a colour version of this figure, please see the plate section.

it can be mistaken to imply that all the information necessary for the construction of an organism lies in the DNA. This is clearly not true. The complete cell is also required, and its complex structures are inherited by self-templating. The 'blueprint', therefore, is the cell as a whole. But that destroys the whole idea of the genome being the full specification. It also blurs and largely nullifies the distinction between replicator and vehicle in selfish gene theory.

'Book of Life'

The genome is often described as the 'book of life'. This was one of the colourful metaphors used when projecting the idea of sequencing the complete human genome. It was a brilliant public relations move. Who could not be intrigued by reading the 'book of life' and unravelling its secrets? And who could resist the promise that, within about a decade, that book would reveal how to treat cancer, heart disease, nervous diseases and diabetes, with a new era of pharmaceutical targets. As we all know, it didn't happen. An editorial in *Nature* in 2010 (which inspired a similar editorial in *Prospect*)[41] spelt this out:

> But for all the intellectual ferment of the past decade, has human health truly benefited from the sequencing of the human genome? A startlingly honest response can be found on pages 674 and 676, where the leaders of the public and private efforts, Francis Collins and Craig Venter, both say 'not much'.

The 'book of life' represents the high watermark of the enthusiasm with which the language of Neo-Darwinism was developed. Its failure to deliver the promised advances in healthcare speaks volumes.[42] Of course, there were very good scientific reasons for sequencing whole genomes. The benefits to evolutionary and comparative biology in particular have been immense, and the sequencing of genomes may well eventually contribute to healthcare when the sequences can be better understood in the context of other essential aspects of physiological function. But the promise of a peep into the 'book of life' leading to a cure for all diseases was a mistake.

The Language of Neo-Darwinism as a Whole

All parts of the Neo-Darwinist forms of representation encourage the use and acceptance of the other parts. Once one accepts the idea that the DNA and RNA templates form a 'code', the idea of the 'genetic programme' follows naturally. That leads on to statements like 'they [genes] created us body and mind', which gets causality wrong in two ways. First, it represents genes as active causes, whereas they are passive templates. Second, it ignores the many feedbacks onto the genome that contribute to circular causality, in which causation runs in both directions. Those mistakes automatically lead to the distinction between replicators and vehicles. The problem begins in accepting the first step, the idea that there is a 'code' forming a complete programme.

The distinction between the replicator and the vehicle can be seen as the culmination of the Neo-Darwinist way of thinking. If all the algorithms for the processes of life lie in the genome, then the rest of the organism does seem to be a disposable vehicle. Only the genome needs to replicate, leaving any old vehicle to carry it.

The distinction, however, is a linguistic confusion and it is incorrect experimentally. The DNA passed on from one generation to the next is based on copies (though not always perfect). The cell that carries the DNA is also a copy (also not always perfect). In order for a cell to give rise to daughter cells, both the DNA and the cell have to be copied. The only difference between copying a cell and copying DNA is that the cell copies itself by growing (copying its own detailed structure gradually, which is an example of self-templating) and then dividing so that each daughter cell has a full complement of the complex cell machinery and its organelles, whereas copying DNA for the purpose of inheritance occurs only when the cell is dividing. Moreover, the complexity of the structure in each case is comparable. When we calculate the amount of information contained in the three-dimensional structure of a cell, it is easy to represent it as containing as much information as the genome. Faithful genome replication also depends on the prior ability of the cell to replicate itself since it is the cell that contains the necessary structures and processes to enable errors in DNA replication to be corrected. The process of self-templating must have developed prior to the development of the relevant DNA.

My germ line cells are therefore just as much 'immortal' (or not) as their DNA. Moreover, nearly all of my cells and DNA die with me. Those that do survive, which are the germ cells and DNA that help to form the next generation, do not do so separately. DNA does not work without a cell. It is simply an incorrect playing with words to single the DNA out as uniquely immortal.

I was also playing with words when I wrote in 2006 in *The Music of Life* that 'DNA alone is inert, dead.' But at least that has a point in actual experiments. DNA alone does nothing. By contrast, cells can continue to function for some time without DNA. Some cells do that naturally, e.g. red blood cells, which live for about 100 days without DNA. Others, such as isolated nerve axons, fibroblasts or any other enucleated cell type, can do so in physiological experiments.

GenesM are best viewed therefore as causes in a passive sense. They do nothing until activated,[43] just as the note on the musical score is not a sound until you hit the keyboard or pluck the string or bang the drum. Active causation lies with proteins, membranes, metabolites, organelles, etc. and the dynamic functional networks they form in interaction with the environment.

Notice also that the language as a whole is strongly anthropomorphic. This is strange, given that most Neo-Darwinists would surely wish to avoid anthropomorphising scientific discovery.

Conclusions

The conclusions of this chapter are twofold.

First, the Neo-Darwinist mechanism involving blind chance variations (copy errors, radiation damage, etc.) followed by natural selection is far from a complete description of what has been found experimentally. We will consider the experimental evidence in more detail in Chapters 7 and 8, where we will also examine the extent to which Neo-Darwinism itself has been tested experimentally.

Second, the language of Neo-Darwinism is itself a powerful barrier to the development of a more inclusive theory. Chapters 7 and 8 will explore more inclusive theories. But first we must state the main principles of Biological Relativity, which is the subject of the next chapter.

Considering the many conceptual problems created by the language of Neo-Darwinism, I even thought about avoiding words like 'gene' and 'code' in this book. Some biologists already do that.[44] But these words are now so deeply ingrained in biology that it can be left to future generations to decide whether to change. I continue to use them while emphasising their limitations.[45]

Notes

1 After finishing *The Origin of Species*, Darwin wrote to Charles Lyell: 'I suppose that I am a very slow thinker, for you would be surprised at the number of years it took me to see clearly what some of the problems were, which had to be solved, – such as the necessity of the principle of divergence of character – the extinction of intermediate varieties on a continuous area with graduated conditions, – the double problem of sterile first crosses & sterile hybrids, &c &c. –'

2 Wallace knew of Darwin's work, and sent his essay on natural selection to Darwin before its publication. The main difference

between them was that 'Wallace was more strictly selectionist than Darwin, who allowed a role for other causes of change'. See http://wallace-online.org/Wallace-Bio-Sketch_John_van_Wyhe .html. Not surprisingly, therefore, Wallace became a Neo-Darwinist. It is not conceivable that the Darwin of *The Origin of Species* and *The Variation of Animals and Plants Under Domestication* could have done so. Those books explicitly acknowledge Lamarckian mechanisms and the second one even proposes a theory that is remarkably similar to modern evidence on transgenerational epigenetics.

3 (20 August 1858) *Journal of the Proceedings of the Linnean Society of London: Zoology* 3:45–50.

4 It is often said that Darwin took a long time to publish *The Origin of Species* because he was afraid of the outcry it would create and only published when it became necessary because of Wallace's work. The historical evidence is not consistent with this idea. See van Wyhe, John (2007) Mind the gap: did Darwin avoid publishing his theory for many years? *Notes and Records of the Royal Society* 61:177–205. A rather different interpretation can be found in Richards, R.J. (2015) Myth 11: that Darwin worked on his theory in secret for twenty years, his fears causing him to delay publication. In *Newton's Apple and Other Myths About Science*. R.L. Numbers and K. Kampourakis, editors (Harvard University Press, Cambridge, MA; pp. 88–95). While agreeing that he was not 'paralyzed by fear' he was cautious to delay publication to be more sure of his ground.

5 The preface to the fourth edition is entitled 'An Historical Sketch of the Recent Progress of Opinion on *The Origin of Species*'. There is also reference to Darwin's grandfather, Erasmus Darwin: 'It is curious how largely my grandfather, Dr Erasmus Darwin, anticipated the views and erroneous grounds of opinion of Lamarck in his "Zoonomia"'.

6 See also the glossary entry on Lamarckism.

7 This kind of inheritance commonly occurs in plants when a sport arises. A sport is a new morphology that is displayed by part of a plant, which can be changes in colour, flowers, leaves or branching. New plants with the changed morphology can be created from cuttings. The morphology is inherited, although sometimes there is reversion to the original form. The new form is therefore not

necessarily stable. This is characteristic of epigenetic inheritance in general. But sometimes the changes are permanent and can be the basis of speciation.

8 The gemmule theory of pangenesis is found in Darwin's 1868 book *The Variation of Animals and Plants Under Domestication*.

9 A remarkable example of such a parallel to Darwin's gemmule idea comes from the work of Corrado Spadafora's laboratory. Cossetti, C. (2014) Soma-to-germline transmission of RNA in mice xenografted with human tumour cells: possible transport by exosomes. *PLoS ONE* 9(7):e101629: 'These results indicate that somatic RNA is transferred to sperm cells, which can therefore act as the final recipients of somatic cell-derived information.'

10 A good example is in a letter that Darwin sent to Moritz Wagner in 1876: 'In my opinion, the greatest error which I have committed, has not been allowing sufficient weight to the direct action of the environment, i.e. food, climate, etc., independently of natural selection.' It is significant that this was written later than the preface to the fourth edition of *The Origin of Species*. His doubts seem to have increased with time. Note that he refers to the *direct* action of the environment, *independently* of natural selection.

11 The emphasis is mine.

12 Romanes actually introduced the term to distance himself from Neo-Darwinism!

13 This mistake was inherent in Weismann's reactions to the nineteenth-century Neo-Lamarckians. His response (1893) to Herbert Spencer, for example, was entitled *Die Allmacht der Naturzuchtung* (the all-sufficiency of natural selection). He wrote 'We accept it [*Allmacht*] . . . simply because we must, because it is the only plausible explanation that we can conceive.' He admitted that it was not possible to observe the process in detail, so there could be no experimental proof, but continued: 'It does not matter whether I am able to do so or not, or whether I could do it well or ill; once it is established that natural selection is the only principle which has to be considered, it necessarily follows that the facts can be correctly explained by natural selection.' For further details on the history of these nineteenth-century debates, see Gould, Stephen Jay (2002) *The Structure of Evolutionary Theory* (Belknap Press of Harvard University Press, Cambridge, MA; pp. 170–250).

14 The precise meaning of the Central Dogma has been much debated. See http://en.wikipedia.org/wiki/Central_dogma_of_molecular_ biology. See also Chapter 7 of this book, where I list the main misinterpretations of the Central Dogma.

15 This kind of dogmatism still flourishes on the internet. There are many blogsites that ridicule, and even libellously insult, those who dissent from standard evolutionary theory.

16 In his book *Evolutionary Genetics*, John Maynard Smith admits (p. 9) 'it is not clear why he thought it [Weismann's claim that the germ line is independent of the soma] was true'. Smith, John Maynard (1998) *Evolutionary Genetics*, 2nd edition (Oxford University Press, Oxford).

17 See Weismann, August (1889) *Essays Upon Heredity* (Clarendon Press, Oxford).

18 Recent research has, however, shown that this is not strictly true. Many genes are involved in eye colour and there can be intermediate colours: www.sciencedaily.com/releases/2007/02/070222180729.htm.

19 Mendel himself showed, however, that this ratio does not always hold. Later (1902) studies by Walter Weldon also showed that there could be a continuous variation in pea colour and a continuous variation in pea shape. See Kampourakis, K. (2015) Myth 16: that Gregor Mendel was a lonely pioneer of genetics, being ahead of his time. In *Newton's Apple and Other Myths About Science*. R.L. Numbers and K. Kampourakis, editors (Harvard University Press, Cambridge, MA; pp. 129–138): 'It thus appeared that in obtaining purebred plants for his experiments, Mendel actually eliminated all natural variation in peas.' An echo of this important point will be found in my description of Conrad Waddington's experiments in Chapter 8. Waddington correctly noted the difference between experiments performed on wild populations and inbred lines.

20 R.A. Fisher in 1936 notably questioned Mendel's accuracy in reporting his results. A careful historical analysis of this issue was published by Fairbanks, D.J. and B. Rytting (2001) Mendelian controversies: a botanical and historical review. *American Journal of Botany* 88:737–752. They conclude that Mendel correctly described his results.

21 Stephen J. Gould documented this process of accommodating exceptions in 1993. Gould, Stephen Jay (1993) Evolution of

organisms. In *The Logic of Life*. C. Boyd and D. Noble, editors (Oxford University Press, Oxford; pp. 15–42).

22 See the exchange of correspondence in 2015. *Journal of Experimental Biology* 218:2658–2659.

23 For a recent review of this process, see Masel, J. (2011) Genetic drift. *Current Biology* 21:R837–R838. The relevant equations of population genetics were developed by Moran, Patrick (1958) Random processes in genetics. *Mathematical Proceedings of the Cambridge Philosophical Society* 54:60–71. See also Moran, P. (1962) *The Statistical Processes of Evolutionary Theory* (Clarendon Press, Oxford).

24 Kimura, M. (1968) Evolutionary rate at the molecular level. *Nature* 217:624–626. Kimura, M. (1983) *The Neutral Theory of Molecular Evolution* (Cambridge University Press, Cambridge). Until Kimura's work, the consensus was that genetic drift would only be significant in small populations through which evolution encountered a kind of bottleneck.

25 West-Eberhard, M.J. (2003) *Developmental Plasticity and Evolution* (Oxford University Press, Oxford).

26 Schrödinger wrote (*What is Life?* p. 101): 'We seem to arrive at the ridiculous conclusion that the clue to understanding of life is that it is based on a pure mechanism, a 'clock-work' in the sense of Planck's paper' (he is referring to Planck, M. (1917) Dynamische und statistische Gesetzmäßigkeit. *Zeitschrift fur Elektrochemie und angewandte physikalische Chemie* 23:63). He continues 'The conclusion is not ridiculous and is, in my opinion, not entirely wrong, but it has to be taken "with a very big grain of salt"'. He then explains the 'big grain of salt' by showing that even clock-work is, 'after all statistical' (p. 103). My reading of these last pages of Schrödinger's book is that he realises that something is not quite right but is struggling to identify what it might be. We would now say that the molecules involved (DNA) *are* subject to statistical variation (copying errors, chemical and radiation damage, etc.), which are then corrected by the protein machinery that enables DNA to be a highly reproducible molecule. This is a three-stage process that reduces the error rate from 1 in 10^4 to around 1 in 10^{10}, which is an astonishing degree of accuracy. The order at the molecular scale is therefore actually created by the system as a whole. This requires energy, of course, which Schrödinger called negative entropy. Perhaps therefore

this is what Schrödinger was struggling towards, but we can only see this more clearly in retrospect. He could not have known how much the genetic molecular material experiences stochasticity and is constrained to be highly reproducible by the organism itself.

27 Both Crick and Watson acknowledged the great influence of Schrödinger's *What is Life?*. See Moore, W. (1989) *Schrödinger: Life and Thought* (Cambridge University Press, Cambridge; pp 403–404).

28 Eldredge and Gould expressed this by saying that the Modern Synthesis became hardened: Eldredge, N. and S.J. Gould (1972) Punctuated equilibria: an alternative to phyletic gradualism. In *Models in Paleobiology*, T.J.M. Schopf, editor (Freeman, Cooper and Co., San Francisco, CA; pp. 82–115). See also Gould, S.J. (1993) Evolution of organisms, which was based on an earlier article: Gould, S.J. (1980) Is a new and general theory of evolution emerging *Paleobiology* 6:119–130.

29 This section of the chapter was first published in Noble, D. (2015) Evolution beyond neo-Darwinism: a new conceptual framework. *Journal of Experimental Biology*, 218:7–13.

30 As an example consider the following two paragraphs:

> Now they swarm in huge colonies, safe inside gigantic lumbering robots, sealed off from the outside world, communicating with it by tortuous indirect routes, manipulating it by remote control. They are in you and me; they created us body and mind; and their preservation is the ultimate rationale for our existence. (*The Selfish Gene*, 1976, p. 21)

> Now they are trapped in huge colonies, locked inside highly intelligent beings, molded by the outside world, communicating with it by complex processes, through which, blindly, as if by magic, function emerges. They are in you and me; we are the system that allows their code to be read; and their preservation is totally dependent on the joy we experience in reproducing ourselves. We are the ultimate rationale for their existence. (*The Music of Life*, 2006, p. 12)

Apart from the obviously true statement 'they are in you and me' the statements are diametrically opposed. Yet no conceivable experiment could distinguish between them. See Noble, D. (2011)

Neo-Darwinism, the Modern Synthesis, and selfish genes: are they of use in physiology? *Journal of Physiology* 589:1007–1015.

31 Keller, Evelyn Fox (2000) *The Century of the Gene* (Harvard University Press, Cambridge, MA): 'As we listen to the ways in which the term is now used by working biologists, we find that the gene has become many things – no longer a single entity but a word with great plasticity, defined only by the specific experimental context in which it is used' (p. 69).

32 The problems arising from conflating different definitions of 'gene' have also been noted in Moss, L. (2004) *What Genes Can't Do* (MIT Press, Cambridge, MA).

33 This diagram needs to be understood in the context of the text explaining the different definitions of a gene and the role of networks in buffering genetic changes. The distinction between phenotype and DNA is itself problematic. DNA is part of the phenotype (McClintock: 'The genome is an organ of the cell').

34 *Stanford Encyclopedia of Philosophy*. Also John Dupré, who advocates 'an atheoretical pluralism' that abandons any pretence to a 'theoretical core to the concept': simply, 'a gene is any bit of DNA that anyone has reason to name and keep track of' (pp. 332–333). Dupré, John (2004) Understanding contemporary genomics. *Perspectives on Science* 12:320–338. http://plato.stanford.edu/entries/human-genome/#HumGenPro.

35 In Dawkins, R. (1982) *The Extended Phenotype* (Freeman, San Francisco, CA).

36 At the time of writing (2015) I believe it is correct to say that most professional biologists no longer regard the version of evolutionary theory presented in *The Selfish Gene* as sufficient or, in its popular version, even relevant. The metaphor, and indeed the whole book, continues however to be deeply influential in other fields, particularly in the social sciences and humanities. It is still therefore necessary to explain what is wrong with it in biological science. My analysis complements that of Midgley, Mary (2010) *The Solitary Self: Darwin and the Selfish Gene* (Acumen, Durham).

37 Page 2 of *The Extended Phenotype*.

38 Noble, Neo-Darwinism, the Modern Synthesis and selfish genes.

39 Wilkins, A.S. (1986) *Genetic Analysis of Animal Development* (Wiley and Sons, New York; p. 485). Wilkins also notes that three of the

architects of the molecular biological revolution, Stent (1980), Brenner (1981) and Jacob (1982), have questioned the validity of the programme metaphor in development.

40 Sun Y.H., S.P. Chen, Y.P. Wang, W. Hu and Z.Y. Zhu. (2005) Cytoplasmic impact on cross-genus cloned fish derived from transgenic common carp (*Cyprinus carpio*) nuclei and goldfish (*Carassius auratus*) enucleated eggs. *Biology of Reproduction* 72:510–515.

41 See the editorial in *Nature*: (2010) The human genome at ten. *Nature* 464:649–650. Also see: Joyner M.J. and F.G. Prendergast (2014) Chasing Mendel: five questions for personalized medicine. *Journal of Physiology* 592:2381–2388.

42 See, for example, Clayton, D.G. (2009) Prediction and interaction in complex disease genetics: experience in type 1 diabetes. *PLoS Genetics* 5:e1000540

43 Genes as DNA sequences are controlled by many switches, which in turn depend on many complex cell, tissue and organ properties. See Ptashne, Mark (2004) *A Genetic Switch*, third edition (CSH Laboratory Press, Cambridge, MA).

44 Examples include Shapiro, James (2011) *Evolution: A View from the Twenty-First Century* (FT Science Press, Upper Saddle River, NJ). Newman, Stuart (2013) The demise of the gene. *Capitalism Nature Socialism* 24:62–72. Philip Kitcher: 'it is hard to see what would be lost by dropping talk of genes from molecular biology and simply discussing the properties of various interesting regions of nucleic acid' (p. 130). Kitcher, P. (1992) Gene: current usages. In *Keywords in Evolutionary Biology*, E.F. Keller and E.A. Lloyd, editors (Harvard University Press, Cambridge, MA pp. 128–131).

45 There is no fully agreed definition of 'gene'. The important point here is that the different definitions of a gene have their strong points and can each be useful in the context appropriate to them. But we should not conflate or confuse them since their consequences are incompatible.

6

Biological Relativity

*He would be unable to determine, how all the parts are modified by the
general nature of blood, and are compelled by it to adapt themselves.*[1]
Benedict de Spinoza, 1665
Letter to Henry Oldenburg, first Secretary of the Royal Society

We arrive now at the point in our story at which the central principle of this book will be laid out. The principle of Biological Relativity is simply that there is no privileged level of causation in biology: living organisms are multi-level open stochastic systems in which the behaviour at any level depends on higher and lower levels and cannot be fully understood in isolation. Just as Special Relativity and General Relativity can be succinctly phrased by saying that there is no global (privileged) frame of reference, Biological Relativity can be phrased as saying that there is no global frame of causality in organisms.

Darwin made the first step in this approach when he proposed that speciation could only be understood in terms of the complete ecosystem (to use modern terminology).[2] The principle of Biological Relativity takes this view to its logical conclusions. One of the mistakes made in Neo-Darwinism was to isolate one part of the 'molecular' level as though it were the whole story, rather than an important part.

Although I will often illustrate the consequences of the principle of Biological Relativity by singling out the molecular, gene-centric form of reductionism (which is the most prevalent form of reductionism), it is important to note at the outset that any form of unwarranted explanatory reductionism is the chief intellectual problem addressed by the theory of Biological Relativity. As an example, the spiral arrhythmia in the heart, discussed in Chapter 3, will appear random and inexplicable at the level of single cells. The order and explanation exist only at the level of the whole organ. In this case the lower level of cells is already one at which the events appear random, without having to go further down to the molecular level.

But why should I be taking you on this journey? What are my credentials for doing so?[3]

A Personal Journey

I have often been asked how I arrived at an integrative and relativistic view of biology, particularly since I started my research career investigating protein ion channels in excitable cell membranes, which was clearly viewed as a reductionist project in the context of more systems-oriented physiology. I come therefore from the reductionist camp. I know and have experienced and been excited by its great strengths. But, for that same reason, I also know its limitations from the inside.

This chapter therefore begins with a story from my own student days at University College London. I went there in 1955 from a London working class tailoring family and was the first member of my family ever to go to university. In those days only 5% of the population benefited from such privilege. It felt like a privilege too. UCL was packed with household names in the various sciences, philosophy, mathematics and many other disciplines.[4] The successes of the reductionist approach were also very apparent. I remember seeing the first images using electron microscopy from Hugh Huxley's work showing the individual protein molecular filaments responsible for muscle contraction. And I could witness the excitement of Bernard Katz's work showing the quantal nature of neuro-muscular transmission.

Sadly, just as I was beginning to push my family's experience to such heights of intellectual endeavour, we suddenly became much poorer through the relatively early death of my father. We became a single-parent family, and there were three younger brothers to bring up. Paying the family's electricity bills from what remained of a student grant was not easy. I shared with our mother the weekly anxieties of managing on very tight budgets. I was not to know then that the youngest brother, Ray, would many years later become one of the most charismatic members of the distinguished faculty at UCL. The significance of that outcome for this book will become clearer in the last chapter. In fact, it infuses the whole of this book since many of the ideas arise from discussions he and I have had over many years.

It was impossible also after that tragic event to imagine how my own career would eventually develop, and even whether I could continue as a student at UCL. When you are trying to make ends meet financially you don't see much beyond the end of each week. The hardship, though, had one very beneficial effect for me. My life had to focus entirely around the family, the academic work and the long cycling journeys across London between the two. With no money left over, student social life became an unnecessary luxury, but therefore no longer a distraction.

It was therefore very fortunate that I won a prestigious Bayliss-Starling scholarship providing the full funding to go on to do graduate work towards a doctorate. By then also our mother had established herself as a tailor in her own right and was finding work. It became possible for me to branch out during my graduate research years. I explored what UCL could offer in physics, in mathematics and in philosophy. Often enough I was the only biologist in these classes. There was no specifically designed course for such an omnivorously hungry student. I simply made it up as I went along, and somehow UCL gave me the freedom to do that. This led to me being the only biologist in the whole university to beg for time on a precious early valve computer: one of the first machines used for seriously challenging mathematics, the Ferranti Mercury. They were so expensive that I doubt whether Ferranti ever made more than a dozen or so of those machines worldwide. I was given the worst time slot each day: 2–4 a.m.! Somehow, in between doing experiments on the heart during the day to obtain the data, I used the evenings and early mornings to develop the mathematical skills to do computer modelling in biology. This experience played havoc with my circadian rhythms and looking back on it I don't know how it was possible to do that for days on end before crashing out for a long weekend sleep.[5]

Every part of this brief autobiographical sketch has played a role in the development of the idea of Biological Relativity, starting with the mathematics.

Ultimate Reductionism: Mathematics?[6]

In spite of the privations, I was happy. I was doing what my professors had taught me, which was that the destiny of biology was to be reduced successfully to chemistry, physics and mathematics. Science was seen as a one-way hierarchy. Each was to be explained in terms of the discipline below it, with mathematics being the supreme goal of this reduction. I was also following the challenge of the great nineteenth-century physiologist Claude Bernard, who as early as 1865 had written '[The] application of mathematics to natural phenomena is the aim of all science, because the expression of the laws of phenomena should always be mathematical.'

As long rolls of computer-generated tapes of results emerged from the Ferranti Mercury, I converted them into beautiful graphs displaying the possible mechanism of heart rhythm. Those graphs found their way into the pages of articles in *Nature* as I experienced directly the excitement of what I believed to be reductionist science. By moving all the way to mathematical analysis, I thought I had become even more reductionist than my teachers! I even started to write on determinism for a student journal at UCL. One of my professors liked the first article so much that he encouraged me to pursue a second one. I had to disappoint him. It never appeared. My thesis research, though, was going very well and was soon to result in two *Nature* publications even before submitting the thesis.

But what was happening was not at all what I imagined it to be. An experience that was aimed to be the ultimate form of reduction had a completely contrary twist in its tail. It was moving me, initially imperceptibly, in precisely the opposite direction. I don't go along much with the idea of eureka ('I've got it!') moments. Switches from one paradigm to another happen more as a consequence of an accumulation of many insights coming together rather than as a single Pauline conversion. But I can point to two key experiences as a research student that paved the way forward (Figure 6.1).

The first was the realisation that the Hodgkin Cycle which I was using in modelling cardiac cells, just as Alan Hodgkin first did with Andrew Huxley for nerve cells, was an example of circular causality, which necessarily involves downward causation from the level of the whole cell to influence the behaviour of its molecules, just as much as upward

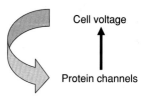

Figure 6.1 The Hodgkin Cycle. Protein channels generate ion movements which determine the voltage of the cell. The voltage in turn controls the opening and closing of the protein channels, and therefore forms part of a feedback system, a form of circular causality. Reproduced with permission.

causation from the molecular level. The more I thought about that the more I realised that, of course, such multi-scale interactions occur everywhere in biology. All physiological functions depend on such circularity. Again, Claude Bernard had led the way with his concept of control of the internal environment in living organisms. This was exhilarating but also rather disturbing. Remember that this period was also witnessing the heyday of molecular biology, with strongly reductionist concepts of genetic programmes and the Central Dogma dominating biological thought. There were also the early signs of intimidation. Those working on a systems approach were openly denigrated as not doing 'real science', not being 'where it is at'. Later, in the 1970s when I became a member of research grant committees, I was to hear that phrase often. Being 'where it is at' was committee-speak for excluding any other approach. Sadly that exclusion was so successful that very little integrative research remained. Molecular biology and genomics sucked up most of the funding.

Introduction to Spinoza

The second experience arose in a somewhat embarrassing way. I was attending the graduate seminars in philosophy run by the Grote Professor, Stuart Hampshire. I am not sure whether he ever knew that I was a gate-crasher, not really one of his own students. If he did know, he was certainly making no concessions to the only non-philosopher in his class. At the end of one of the seminars he singled me out to introduce the next seminar. Each of the students was expected to do this in turn. In a panic, and with just a week to do it, I had to choose a topic and write a ten-minute plausible introduction. There was no time to plough through dense philosophical texts in the library. Anyway, there were experiments to perform during the day and there was my assignment with Mercury at night. I had to do what we call winging it, flying by the seat of one's pants. I raided my premature scribblings on determinism to see what I could come up with. I decided to give an account of freewill from a physiologist's perspective by distinguishing causes internal to the organism from causes external to it. The idea was that only the former could be the basis of free action.

If you have absorbed the central message of this book, you will already see the naivety in this approach. It is in the *interactions* between the two that we should be looking for clues. But I did not see it that way then. Hampshire's response showed that he wasn't buying it: '*Very* interesting, Mr Noble.' Well, at least he knew my name even if it wasn't on his student list! I have also become wary of professional philosophers who begin their comments with '*very* interesting', a sure sign that something is wrong. He continued: 'Mr Noble, you should read Spinoza.' That was it. End of story. The rest of the seminar has become a haze in my memory as the class dissected the paper and pulled it to pieces. All I can now recall is finding myself at the local bookshop to buy Hampshire's excellent book, *Spinoza*,[7] and an edition of the collected works and letters of Spinoza. I still have both of them. Hampshire's book was beautifully clear. The collected works were dense and seemed incomprehensible. But I now know that I was looking in the wrong place. I should have focused on the letters at the end of the book. I didn't read those until 40 years later.

Spinoza's Way Out of the Cartesian Paradigm

In Western scientific thought the mechanistic reductionist view can be seen to have originated with Descartes in the seventeenth century, while relying heavily on Newtonian mechanics later in the century, and in later centuries on the mathematical genius of Pierre-Simon Laplace. The philosopher René Descartes laid the foundation by arguing that animals could be regarded as machines in some way comparable to the ingenious

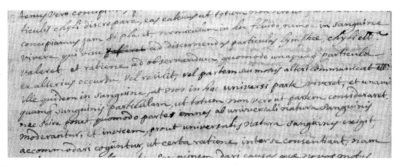

Figure 6.2 Part of Spinoza's letter to the Royal Society Secretary, 1665. The Latin text of the section translated into English here begins 'concipiamus jam, si placet...' ('Let us imagine, with your permission...').

hydrostatic robots that had become popular amongst the aristocracy in their gardens. Newtonian mechanics cemented the foundation with the laws of mechanical motion, and Laplace systematised the ideas with his famous statement that a supreme intelligence could use mathematics to predict the future completely, and retrodict the past as well. Everything that has or will happen would be clear to such a being.

Descartes even foresaw one of the central ideas of Neo-Darwinism:

If one had a proper knowledge of all the parts of the semen of some species of animal in particular, for example of man, one might be able to deduce the whole form and configuration of each of its members from this alone, by means of entirely mathematical and certain arguments, the complete figure and the conformation of its members.[8]

This is essentially the idea that there is a complete mathematical 'programme' there in the semen, prefiguring Jacob and Monod's 'genetic programme'. Complete because he writes 'from this alone'.[9]

It is therefore very significant that the earliest traces of the concept of Biological Relativity can be traced back to Descartes' main philosophical opponent. In 1665, just two years after the foundation of the Royal Society, Benedict de Spinoza, working in Holland, was in extensive correspondence with the first Secretary of that Society, Henry Oldenburg, working in London (Figure 6.2).

Oldenburg had just returned from meeting Spinoza in Holland and had been fascinated by his discussions with him on 'the principles of the Cartesian and Baconian philosophies'. Spinoza was opposed to the dualism of mind and body espoused by Descartes. This was necessary in Descartes' view of animals as automata since he wished to exclude humans from this view and so attributed their freewill to a separate substance, the soul, which could interact with the body. Spinoza was in the process of seeking to publish his great work (The ethics: *Ethica ordine geometrico demonstrata*) in which he proposes an alternative philosophy. The correspondence shows how close the *Philosophical Transactions* of the Royal Society came to being the vehicle for one of his seminal publications. In the end, however, Spinoza did not publish in *Philosophical Transactions*, but this correspondence includes an important letter from Spinoza which could form a text for the systems approach and for one of the bases for the concept of Biological Relativity. The original letter in Latin is still kept in the Royal Society library. He writes: 'every part of nature agrees with the whole, and is associated with all other parts' and 'by the association of parts, then, I merely mean that the laws or nature of one part adapt themselves to the laws or nature of another part, so as to cause the least possible inconsistency'. He realised therefore some of the problems faced in trying to understand what, today, we would characterise as an open system.

In particular, he appreciated the difficulty in working from knowledge of minute components to an understanding of the whole:

> Let us imagine, with your permission, a little worm, living in the blood, able to distinguish by sight the particles of blood, lymph etc, and to reflect on the manner in which each particle, on meeting with another particle, either is repulsed, or communicates a portion of its own motion. This little worm would live in the blood, in the same way as we live in a part of the universe, and would consider each particle of blood, not as a part, but as a whole. He would be unable to determine, how all the parts are modified by the general nature of blood, and are compelled by it to adapt themselves, so as to stand in a fixed relation to one another.

This paragraph could stand even today as a succinct statement of one of the main ideas of Biological Relativity. He doesn't use a mathematical medium to express his idea, but this could be so expressed as the aim to understand how the initial and boundary conditions of a system constrain the parts to produce a particular solution to the differential equations describing their motions. We need, then, to move to the complete system (with whatever boundary we choose to use to define that) in order even to understand the parts.

In this chapter I will express Spinoza's main concept in modern terms. We will see that Spinoza's idea is a necessary but not a sufficient condition to complete the escape from Cartesian determinism. There are two further conditions that need to be satisfied to arrive at the theory of Biological Relativity. These conditions will be described later (see Why Spinoza's Constraint is Not Sufficient).

The Essence of Biological Relativity

To get an immediate practical experience of the main idea, take some knitting needles and some wool. Knit a rectangle. If you don't knit, just imagine the rectangle. Or use an old knitted scarf, preferably a loose knit so that you can see how the individual knots behave. Now pull on one corner of the rectangle while keeping the opposite corner fixed. What happens? The whole network of knitted knots moves. Now reverse the corners and pull on the other corner. Again the whole network moves, though in a different way. This is a property of networks. Everything ultimately connects to everything else. Any part of the network can be the prime mover, and be the cause of the rest of the network moving and adjusting to the tension. Actually, in living systems, it would be better still to drop the idea of any specific element as a prime mover. It is networks that are dynamically functional.

Now knit a three-dimensional network. Again, imagine it. You probably don't actually know how to knit such a thing. Pulling on any part of the three-dimensional structure will cause all other parts to move. It doesn't matter whether you pull on the bottom, the top or the sides. All can be regarded as equivalent. There is no privileged location within the network.

The three-dimensional network recalls Waddington's epigenetic landscape network (Figure 6.3) and is quite a good analogy to biological networks since the third dimension can be viewed as representing the multiscale nature of biological networks. Properties at the scales of cells, tissues

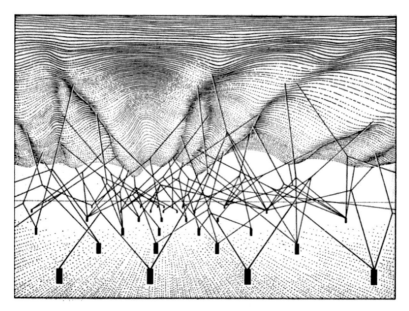

Figure 6.3 Conrad Waddington's diagram of the epigenetic landscape from *The Strategy of the Genes*. Genes (solid pegs at the bottom) are viewed as parts of complex networks so that many gene products interact between themselves and with the phenotype to produce the phenotypic landscape (top) through which development occurs. Waddington's insight was that new forms could arise through new combinations to produce new landscapes in response to environmental pressure, and that these could then be assimilated into the genome (image from Waddington 1957).[10]

and organs influence activities of elements, such as genes and proteins, at the lower scales. This is sometimes called downward causation, to distinguish it from the reductionist interpretation of causation as upward causation.[11] This is the mechanism of the constraints in Spinoza's text quoted above.

'Down' and 'up' here are metaphors and should be treated carefully. The essential point is the more neutral statement: there is no privileged level of causality. This must be the case in organisms which work through many forms of circular causality. One of the consequences of the relativistic view is that genesM (see Chapter 5 for the different definitions of a gene) cease to be represented as active causes.[12] Templates are passive causes, used when needed. Active causation resides in the networks which include many components for which there are no DNA templates. So, genesJ do not behave like genesM since they also include network

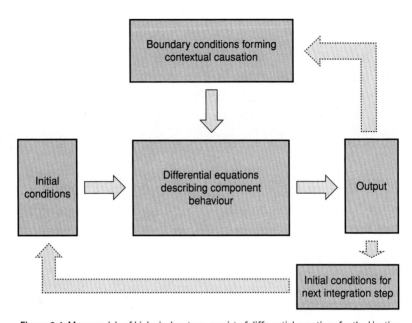

Figure 6.4 Many models of biological systems consist of differential equations for the kinetics of each component. These equations cannot give a solution (the output) without setting the initial conditions (the state of the components at the time at which the simulation begins) and the boundary conditions. The boundary conditions define what constraints are imposed on the system by its environment and can therefore be considered as a form of contextual causation from a higher scale. The arrows are not really unidirectional. The dotted arrows complete the diagram to show that the output contributes to the boundary conditions (although not uniquely), and they determine the initial conditions for the *next* integration step (legend and diagram from Noble, 2012).[13]

properties. It is the physics and chemistry of those dynamic networks that enable them to function, but just as the properties of molecules emerge from the interactions between the component atoms, so the dynamic networks emerge from the interactions of the individual molecules and structural components. The individual components are constrained by being parts of the network (Figure 6.4).

An important linguistic feature of the alternative, relativistic, concepts proposed here is that most or all the anthropomorphic features of the Neo-Darwinist language can be eliminated, without contravening a single biological experimental fact. There may be other forms of representation that can achieve the same result. It doesn't really matter which you use. The aim is simply to distance ourselves from the biased conceptual scheme that Neo-Darwinism has brought to biology, made more problematic by the fact that it has been presented as literal truth.

Conceptual and Empirical Interpretations

The principle that there is no privileged scale of causality can easily be misinterpreted. It is important now to introduce some clarifications.

First, we must distinguish between its conceptual status and its practical implications. It is an a-priori statement, i.e. a statement about what we should or should not assume in advance of doing the experiments. We should not assume that causation *necessarily* resides at a particular, e.g. molecular, scale. That is the mistake made by naive reductionism in biology. The reduction to molecular-level events is treated as a methodological necessity, whereas it should emerge, if at all, from the experiments themselves. Before we do those experiments, we can't know which parts of a system are involved in its behaviour, nor attribute any privileged position to them.

But that does not mean that all scales must be involved in any given example. The circles of causal networks may span particular ranges of scales, which may be more or less limited in extent. And there may be particular levels that act as important hubs. Those facts are for us to discover as empirical observations. For example, many biologists, me included, believe that the cell is a central level of integration in much of biology. That conviction is a result of extensive experiments on cells showing their functional integrity and that many physiological functions cannot be ascribed to entities lower than the cell. Cells contain the main metabolic networks, circadian and various other rhythm networks, cell cycle networks and so on.

The important position of the cell is not surprising since for billions of years all organisms were unicellular. And even in multicellular organisms like you and me, the cell is the level at which we pass on most inherited characteristics. There is a bottleneck in the reproductive process. Eggs and sperm are single cells, and so is their union before embryonic development. All organisms that reproduce sexually necessarily go through a single-cell stage in their life cycles.

The genome also has a unique position. But it is not the one most often ascribed to it as a programme dictating life. As the American cell biochemist Franklin Harold puts it in his book *In Search of Cell History*, 'The genome is not the cell's command center but a highly privileged databank, something like a recipe or a musical score, yet for the purpose of parsing evolution, genes have a rightful claim to center stage.' Parsing is the analysis of strings of symbols, usually with guidance from some rules of grammar. In the case of DNA, the start and stop sequences and those for binding transcription factors, amongst other features, provide those guidelines. Analysis of this kind has indeed been exceptionally useful in the inter-species DNA sequence comparisons that now form the basis of much of our understanding of evolutionary history. This is the method by which Carl Woese's discovery of the archaea was confirmed (see Chapter 4).

Each feature of organisms at the various levels may have unique properties. The principle of Biological Relativity should not be taken to require that all forms of causation involved are equivalent. We will meet the different forms of causation later in this chapter.

With regard to possible boundaries of multi-scale interactions, it is unlikely that scales below that of atoms will be relevant in physiological processes except in the general sense of having the properties that make life possible. For similar reasons, scales above that of the planet are unlikely to be more involved than making life possible.

But, as we have seen, there are many scales between these two extremes. Moreover, it is now known as a matter of fact that interactions between all of the scales involved in organisms and their environment occur. As Spinoza foresaw, the smallest particles, which for us are the molecules of living organisms, are constrained by their cellular environment, which in turn are constrained in multicellular organisms by the physiological properties of tissues, organs and systems, while whole organisms are constrained by their environment, including interactions with other organisms, which means that sociological factors also matter.

If, like Spinoza's imagined little worm, we focus on observing the motions of those smallest particles, we will find that they do indeed obey the laws of physics and chemistry in their interactions with other molecules. In that sense, and in that sense alone, biology can in principle be reduced to physics and chemistry. It is when we focus on more global, but nonetheless real, properties that we can see the limits of that reduction. Even constraining some gas molecules by putting them in a

container constrains their motions so that there will be an overall property, the pressure of the gas, within the container. In any equations we use
to describe what is happening, that constraint will appear in the boundary conditions that must be inserted into our model to enable a solution
to be found. This fact is universal. All models, even entirely determinate ones, require initial and boundary conditions to be inserted into the
equations before we can make any predictions on what may happen.

Open and Closed Systems

Initial and boundary conditions will be required even if we isolate the
container from any other influences beyond its boundaries. That is what
we call a closed system. A closed system, however, is an idealisation for
experimental purposes. Real systems are never strictly isolated. Even a
container would need to be completely rigid (which is an idealisation) to
be strictly isolated. In practice, the container material itself will be subject
to the pressures on it, within and without. That is very obvious indeed if
the container is a balloon, which will readily expand if we put more gas
inside it as the balloon material is elastic. Even the most rigid material,
however, has some degree of elasticity. In physical experiments we simply
reduce that elasticity to the point at which such variations are negligible.
Physicists have to go to extraordinary lengths to achieve isolation of their
experiments from outside influences.

But we cannot possibly do that with living systems. Organisms are
necessarily open systems. They must exchange energy and matter with
their environment in order to survive. They are, therefore, active systems
and they manage this interaction with the environment according to the
control processes that exist within them. The constraints on behaviour
can then take many different forms, depending on how those exchanges
occur and what controls them.

Of course, it is possible to perform experiments on organisms and
their components that aim at the idealisation of a closed system. But it is
then important to understand that we will no longer be experimenting
on a living organism. It will be dead. Many biological experiments nevertheless involve such procedures. Experiments done particularly to determine the structure of the molecules of living organisms are performed in
this way. DNA was not part of a living organism when its double helical

structure was worked out by Watson, Crick, Wilkins and Franklin.[14] There really is nothing alive in the DNA molecule alone. If I could completely isolate a whole genome, put it in a petri dish with as many nutrients as we may wish, I could keep it for 10,000 years and it would do absolutely nothing other than to slowly degrade.

Physiological investigations to determine function and to reveal control mechanisms must, however, also involve living systems. There is then no escape from the constraint that they are open systems.

Why Spinoza's Constraint is not Sufficient

So far in this chapter I have expressed Spinoza's idea of the constraint of parts by the whole in terms of the constraint by a larger scale. What would happen if we extend out to the largest scale of all, the universe? His answer would be that we then arrive back at a completely deterministic system. This is Nature as a whole, which for Spinoza was the prime mover, or God, *Deus sive Natura* (God or Nature).

Two further conditions are required to complete the theory of Biological Relativity and to escape fully from the Cartesian paradigm. The first is stochasticity. The second is the distinction between scales and levels.

Stochasticity. The idea that the genetic material is read in a fully determinate way was originally proposed by Schrödinger in 1943.[15] But it was also implied by Descartes (see quotation earlier in this chapter) and by Weismann's Barrier idea. These authors did not know about DNA, but they assumed that whatever was the genetic material was the sole determinant of the organism's structure and development. This idea is a mistake and its correction is essential to the theory of Biological Relativity.

It is true, of course, that at the micro level there is a determinate readout in the sense that the DNA sequence uniquely determines the protein amino acid sequence, as described in Chapter 5 (Figure 5.2). But this is not sufficient to specify the phenotype. That also depends on the pattern formed by the relative quantities of the different proteins expressed. That pattern is not specified by the DNA sequence. It is specified by the cells, tissues and organs of the body through the control they exert on the expression levels of different proteins. Those levels are also

Figure 1.1 A view of the Milky Way towards the Constellation Sagittarius (including the Galactic Centre) as seen from a non-light polluted area (the Black Rock Desert, Nevada) (courtesy of Steve Jurvetson).

Figure 1.2 Jupiter and the four Galilean moons observed through a Meade "10" LX200 telescope, i.e. ten times more magnification than was available to Galileo (Jan Sandberg, Wikimedia).

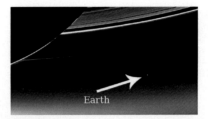

Figure 1.3 The Earth (little blue dot) viewed from the spacecraft *Cassini* as it photographed the rings of Saturn during an eclipse of the sun by Saturn. *Cassini* was 900 million miles from Earth. Light takes over an hour to reach the Earth from Saturn. But this is miniscule compared to the more than 13 billion years for light to reach us from the edge of the observable universe (source: NASA/JPL-Caltech/Space Science Institute (www.nasa.gov/mission_pages/cassini/multimedia/pia17171.html)).

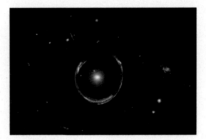

Figure 1.4 A remarkable example of gravitational lensing predicted by the theory of Relativity. The luminous red galaxy in the centre of this image, LRG 3-757, is almost exactly in front of a much more distant blue galaxy whose light is bent by the gravitational 'lens' to produce an almost perfect circle (source: Lensshoe_hubble.jpg: ESA/Hubble & NASA).

Figure 1.5 The Hubble deep field view of one-24-millionth of the sky showing numerous galaxies right to the edge of the visible universe (source: apod.nasa.gov/apod/ap140605.html).

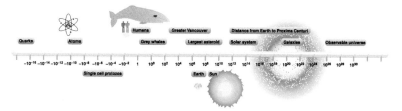

Figure 2.2 Scales from the tiniest subatomic particles to the whole observable universe. On the logarithmic scale used here, man lies roughly midway between the smallest subatomic particles and a whole galaxy (http://hendrix2.uoregon.edu/~imamura/123/lecture-1/lecture-1.html).

Elements in the periodic table occurring in living systems

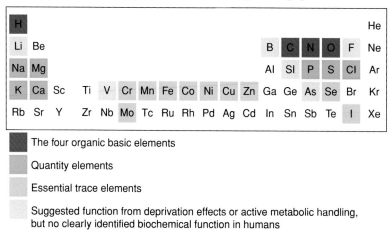

The four organic basic elements

Quantity elements

Essential trace elements

Suggested function from deprivation effects or active metabolic handling, but no clearly identified biochemical function in humans

Figure 2.4 Part of the periodic table showing the elements found in living organisms in the lower-weight region of the table (adapted from: https://en.wikipedia.org/wiki/Template: Periodic_table_(nutritional_elements)).

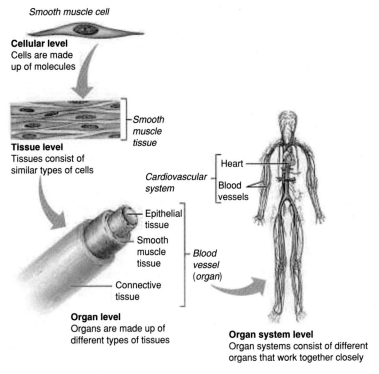

Smooth muscle cell

Cellular level
Cells are made up of molecules

Tissue level
Tissues consist of similar types of cells

Smooth muscle tissue

Cardiovascular system

Heart
Blood vessels

Epithelial tissue
Smooth muscle tissue
Connective tissue

Blood vessel (*organ*)

Organ level
Organs are made up of different types of tissues

Organ system level
Organ systems consist of different organs that work together closely

Figure 2.7 Images of cells, tissues, organs and organ system levels using smooth muscle as an example (*J Physiol.* 2014 Jun 1; 592: 2375–2379).

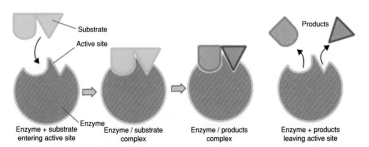

Substrate
Active site

Products

Enzyme + substrate entering active site

Enzyme
Enzyme / substrate complex

Enzyme / products complex

Enzyme + products leaving active site

Figure 3.2 Lock-and-key representation of the way in which enzymes work. In this case the two parts of a molecule to be broken up fit into the template formed by the enzyme, which then influences the molecule to make it easier to break the chemical bonds. The reverse process would describe how an enzyme brings two separate molecules together to make them more likely to join. This is how enzymes enable long strings of nucleic or amino acids to link together to form DNA or proteins (https://commons.wikimedia.org/wiki/File:Two_Substrates.svg).

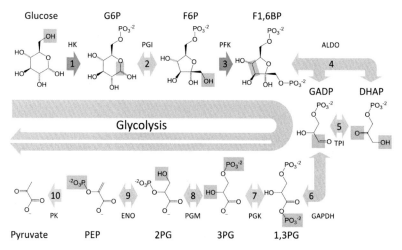

Figure 3.3 Diagram of a 'small' network. This network is called the glycolytic pathway and has ten links, each one of which is enabled by a particular protein enzyme. The links represent the processes by which glucose is broken down to produce energy. There is no need to remember the details of this diagram. The important details can be found on Wikipedia under the entries 'glycolysis' and 'enzyme'. What is relevant to this chapter is that even small processes in biological networks are complex in their details (from Wikimedia Commons, Thomas Shafee).

Figure 3.4 Spiral waves in nature at very different scales: galaxy (left), cyclone (middle) and heart arrhythmia (right). A spiral galaxy extends across tens of thousands of light years; a spiral weather system extends across hundreds of miles; a spiral wave in the heart extends across a few centimetres (sources: left: the pinwheel galaxy Messier 101 (www.spacetelescope.org/images/heic0602a, European Space Agency and NASA); middle: http://visibleearth.nasa.gov, Jacques Descloitres, MODIS Rapid Response Team, NASA/GSFC; right: S. Panfilov/Univ.Ghent).

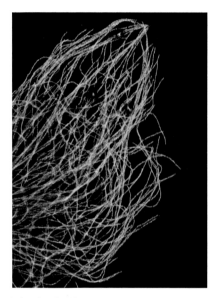

Figure 4.1 Microtubules visualised in a part of a cell. Just as our bodies rely on bones for structural support, cells rely on a cellular skeleton. In addition to helping cells keep their shape, this cytoskeleton transports material within cells, is responsible for cell movement and coordinates cell division. Microtubules are shown here as thin strands. Each tubule is about 24 nm across, but extends over distances than can go from one end of a cell to another, which is many micrometres (from US Department of Health and Human Services (public domain)).

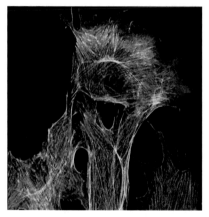

Figure 4.2 Microtubules (green) and actin microfilaments (orange) in an epithelial cell from the lung of a cow. The tubules are about 24 nm across; the filaments about 6 nm across (from US Department of Health and Human Services (public domain)).

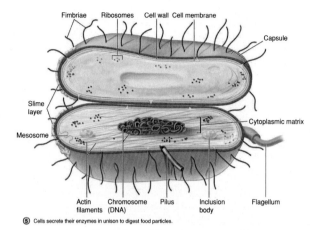

Fimbriae   Ribosomes   Cell wall   Cell membrane

Capsule

Slime layer

Mesosome

Cytoplasmic matrix

Actin filaments   Chromosome (DNA)   Pilus   Inclusion body   Flagellum

⑤ Cells secrete their enzymes in unison to digest food particles.

Figure 4.4 Typical structure of a bacterium. There is no nucleus. The DNA is instead coiled up to form what is called a nucleoid. Various protrusions form structures that can propel the organism (flagella) and transfer DNA between bacteria (pili). A bacterium with this kind of shape (there are many other shapes!) would be about 5–10 μm long and about 1 μm wide (https://commons .wikimedia.org/wiki/File:Major_events_in_mitosis.svg). Copyright McGraw-Hill companies inc.

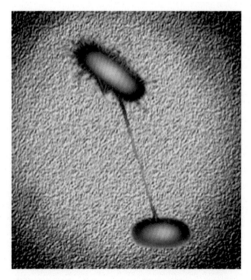

Figure 4.5 Image of the connection between two pili stretching out between the two bacteria. Each bacterium is about 1 μm across. The bridge is about 6 μm long. It contracts to draw the two bacteria close together. Exchange of DNA can then occur (http://image.slidesharecdn.com/).

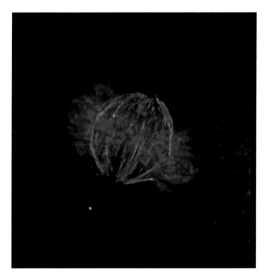

Figure 4.8 Cell in mitosis at the stage of the spindle (green filaments) having formed before the chromosomes (red) are separated and pulled to each end of the cell (Roy van Heesbeen, via Wikimedia Commons).

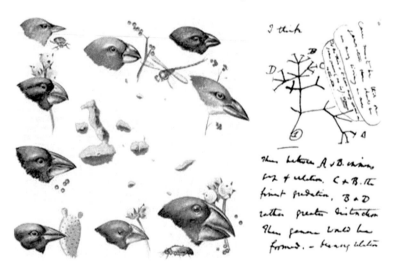

Figure 5.1 Adaptive radiation in the finches of the Galapagos islands, with Darwin's diagram of radiation beginning 'I think...'. He observed similar radiation in the tortoises on the different islands (www.biomedware.com/blog/2012/genetic-gis-a-call-and-a-research-agenda/, right: Charles Darwin's 1837 sketch of an evolutionary tree from his *First Notebook on Transmutation of Species* (1837)).

Figure 5.4 Cross-species clone. The nucleus of a common carp, *Cyprinus carpio* (middle), was transferred into the enucleated egg cell of a goldfish, *Carassius auratus* (left). The result is a cross-species clone (right) with a vertebral number closer to that of a goldfish (26–28) than of a carp (33–36) and with a more rounded body than a carp. The bottom illustrations are X-ray images of the animals in the top illustration (figure kindly provided by Professor Yonghua Sun from the work of Sun *et al.* (2005).

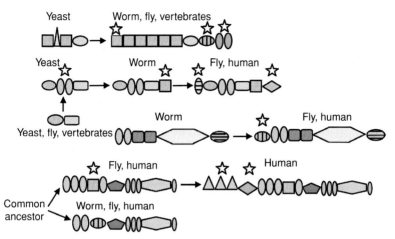

Figure 7.4 Evidence for how gene domains must have moved around during evolution to form different functional proteins by domain accretion. This figure shows diagrammatic representations of three groups of individual proteins that contribute to the formation of chromatin, the backbone of chromosomes. Each coloured shape shows a single protein domain. Each protein is made up of several domains. In each case, as we move from yeast (unicellular) to worm, fly, various vertebrates and the human, the number of domains increases. The significant fact is that the development from one to the other did not occur by gradual accumulation of small mutations. Instead, whole domains hundreds of amino acids in length have come together to form the new protein. The red stars show the domains that must have moved in this way. It is extremely unlikely that this result could have been achieved by gradual accumulation of small random mutations. Reproduced with permission. For full details of the domains and proteins involved, see the original *Nature* paper at www.nature.com/nature/journal/v409/n6822/fig_tab/409860a0_F42.html.

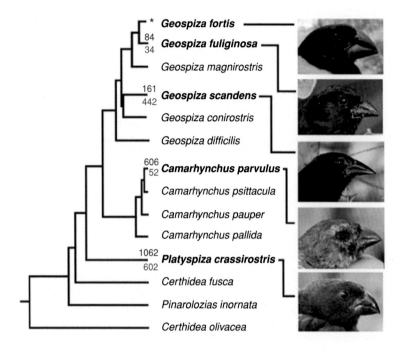

* – Reference species
Red – Epimutations (DMR)
Blue – Genetic mutations (CNV)

Figure 8.2 Genetic and epigenetic variation in the finches of the Galapagos Islands. The tree diagram on the left shows the relatedness of the 14 individual species. Four of these species were compared with the one chosen as the reference species, *Geospiza fortis*. DMR = DNA methylation region; CNV = copy number variation. The number of epigenetic mutations is shown as red, the number of genetic mutations in blue (from Skinner *et al.* 2014).

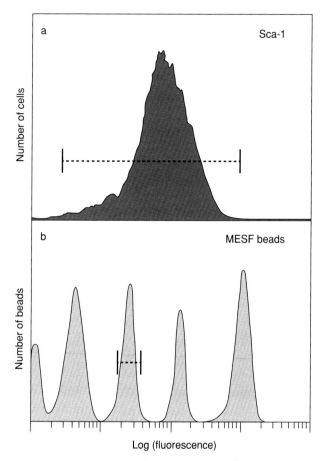

Figure 6.5 Variation in expression levels of a protein[16] in cells within a cultured population. The top image shows the numbers of cells displaying various levels of expression. The range, indicated by the horizontal bar, is 1000-fold, i.e. three orders of magnitude. The lower image shows the resolution of the method by which the levels were measured, which is clearly well within the accuracy required (from Chang *et al.* 2008 [17]).

strongly stochastic. This fact was first observed in biochemical components of organisms by R.J. Williams in 1956,[18] and it has been beautifully confirmed by recent experiments revealing the extent of stochasticity in gene expression. Neighbouring cells in a population can differ in the level of expression by as much as three orders of magnitude (Figure 6.5).

This extensive variation leaves it wide open to an organism to select what works in any given situation. In fact, the same experiments that produced the result in Figure 6.5 also showed that the distribution curve is a property of the population as a whole, not of single cells. In the sense used in this book, that distribution is an attractor created by the population.

Levels and emergent properties. The difference between scales and levels was introduced in Chapter 2. Scale is a matter only of extension, i.e. how large a part of nature is considered. Level is a matter of organisation, which could be a cell, an organ, an organism, a population and so on. In the context of the theory of Biological Relativity, the difference is crucial. A level is a functional entity with goals and natural ability to achieve those goals. The level of a cell, for example, exists because cells are integrated functional entities with properties that emerge as the interactions of their component parts within the constraints exerted by the cell as a whole. We encountered some of those emergent properties in Chapter 3, including circadian and cardiac rhythms. These emergent properties not only characterise what a cell is, they also constrain (in Spinoza's sense) the component, e.g. molecular, parts to conform to the goals of the cell. It is through these meaningful constraints that the behaviour of the whole can be said to display natural purposiveness.

The theory of Biological Relativity requires all three of these necessary conditions: constraint of components by the whole, stochasticity and emergent goal-directed properties. Together they satisfy the requirements for conditioned arising,[19] the processes by which each development creates the conditions for further development, with stochasticity guaranteeing that the processes cannot be completely determinate.[20]

Forms of Causation

The requirement for initial and boundary conditions to solve the equations for any system already introduces us to two different forms of causation. The laws of motion represented by differential equations will be one form which is dynamic, while the conditions will be another form which is more like a formal cause. Once we introduce control processes and the activity of organisms which those processes enable, there will be even more forms to take into account. Philosophers have been familiar with the need to consider different forms of causation ever since Aristotle outlined his four forms over 2000 years ago.[21] Those four forms are still relevant, though they need updating to take account of the kinds of causal interactions that we now know about.

Material cause. This refers to the nature of the material of which something is made. A table, for example, will behave differently dependent on whether it is made from wood, metal, glass, stone, etc. Some materials can burn easily, bend easily, transmit light easily, others cannot. We can update this to include everything we know about the physico-chemical composition of the materials in living systems. It is in this sense that we can say that even the subatomic scale is relevant. It is relevant to what makes it possible for the materials to exist and have the properties they have. Clearly this is a passive cause in the context of understanding biology. In relation to living organisms the material structure of the universe is a cause in this sense.

Formal cause. This refers to the way in which the material is arranged. The same material arranged as a chair will serve a different purpose and be responsible for different things to happen than when the same material is arranged as a table. Today, we would have difficulty drawing the same distinction between the first two Aristotelian forms. The same atoms can be arranged differently to produce materials with very different properties. Notably, carbon atoms can be arranged to produce soft graphite, or to produce super-hard diamond. Aristotle, of course, would not have known about atoms, or our subatomic particles. We can make a pencil out of graphite, but not out of a diamond, which would be used for very different purposes for which a pencil would be useless. The distinction between material and formal cause therefore depends on the scale at which we are working. A difference in formal cause at one scale can become a material cause at a higher scale. Formal causes are very important in biological systems since their organisation is what makes life possible. This is the correct category for genes. The DNA sequences are a form of structure, an arrangement of the DNA molecule, a kind of molecular anatomy.

Efficient cause. This is the active form of causation central to reductionist accounts of living systems, and it refers to what acts to produce dynamic effects, for example one billiard ball hitting another to send it off with new direction and velocity. The mistake in the reductionist account is to assume that there is necessarily and always a prime mover. We need therefore to add the principle of relativity since it is necessary to recognise that in many systems of interactions involving circular causality there will be no 'prime mover'. In the Hodgkin Cycle, for example, it doesn't make much sense to ask whether the protein channels or the cell voltage form the prime mover since they depend on and move each other. This is evident also in the differential equations we use in Hodgkin–Huxley type models. In each time step of the integration both limbs of the cycle are computed together. This will be true in all forms of circular causality. This is where Biological Relativity has one of its most important consequences. Looking for a 'prime mover' is the wrong strategy when we should be focusing on interactions in processes.

Final cause. This is what the action is used for in the sense of its function and its purpose. The heart is for the pumping of blood; the lungs are for the exchange of respiratory gases, oxygen and carbon dioxide; eggs and sperm are for the reproduction of the species; and so on. This is the form of causation that gives pure reductionists most concern and why they usually wish to eliminate it from biological theory. They would argue that teleological cause can always be replaced by explanations in terms of efficient (mechanical) cause. But that approach privileges the level at which causation is thought to occur, usually represented as molecular. There is no reason to make that assumption and in open systems it must be wrong to do so. Functional significance, purpose and therefore teleological causation arise as inter-scale interactions in which the components at the lower scale are constrained in a functional way by the properties at a higher scale. This is the essential point that was understood by Spinoza, as we saw earlier in this chapter.

As we have also seen in this chapter, expressed rigorously in mathematical terms there is nothing mysterious about such constraints. All open systems are subject to them. The causation involved is just as effective as any other form. Scientifically testable theories should therefore incorporate such causation. The problem for some scientists has perhaps arisen from the word 'final'. From a mechanistic viewpoint, such causation is of course relative. The word 'final' does not entail that there is some overall purpose, an intelligent designer or other theological assumptions. It simply means 'purpose'. Those purposes arise from the inter-scale interactions. Processes at higher scales give purpose to processes at lower scales. This is a central feature of the concept of the relativity of epistemology, which will be explored in Chapter 9.

Consider as a concrete example the regularity of the normal heartbeat and how it is disturbed in life-threatening arrhythmias which I described in Chapter 3. The normal heartbeat is an attractor caused by a circular form of causality in which both the cell potential and the individual proteins are entrained by their interaction. Once the rhythm begins it can continue indefinitely. Even large perturbations in the individual proteins or their genes can be resisted. Now consider what happens when a different kind of attractor is established. This happens in the heart when abnormal spiral waves of excitation arise at the level of the whole heart. The individual molecules in each cell are now constrained to dance to a different and much more chaotic rhythm. Viewed from the level of the individual molecules, both of these influences from the higher levels of the cell or the whole organ will seem inexplicable, even apparently random. The molecules are like boats tossed around in a storm beyond their own control. Yet, the storm also depends on their activity. Indeed, it can be modelled using the equations for that activity. Each of the three views – the molecular, the cellular and the organ – are valid, but only the higher levels provide an explanation of what is happening.

Using this example, we can also clarify the distinction between function and purpose. The function of the protein–membrane network that ensures cardiac rhythm is clear at the level of cells, but does not even exist at the level of molecules. Its purpose, though, becomes clear at an even higher level, i.e. the cardiovascular system. A single cell 'knows' nothing about that purpose. The distinction between function and purpose can therefore be a matter of the level at which each is relevant.[22]

Beyond the Aristotelian forms. While it is historically necessary to recognise the insights of Aristotle since his classification of the forms of causation is still very relevant today, it is also important to note that developments in modern science have introduced variations on his theme that he could not have anticipated.

Attractors. The concept of an attractor is a good example. If the functional process expresses itself as an attractor it literally pulls the system towards a goal. Within what is called the basin of the attractor, the components have to go along with it.[23] That is another way of expressing the influence of a higher scale on the behaviour of components at a lower scale. We met examples of these in the cases of heart rhythm and circadian rhythm in Chapter 3. On astronomical scales a black hole acts as an attractor, as does a spiral galaxy.

Coding. The concept of coding also introduces a form of causation that is new, even though it can be represented within Aristotle's classification. Within his system, coding can be seen as a formal cause since it concerns the way in which components are arranged. But I don't think that his idea of formal cause captures the full weight of the concept of coding in molecular biology. There is something special about the discovery of the triplet code for the way in which nucleic acid sequences act as templates for amino acids in protein sequences. Equally, however, we should not endow molecular biological coding with implications beyond what the molecular biological facts show. This is where the concept of coding has been misinterpreted. Molecular biological coding does not have the same meaning as coding in the sense of intentional human communication. That sense of coding occurs in a context in which the messages encoded have meaning. There is no meaning in the nucleic acid sequences themselves. Meaning arises as a consequence of the contextual logic within which they are interpreted within a living organism. On their own, the sequences have no more significance than the bar codes of products. If all that we knew about organisms was restricted to knowledge of their DNA sequences, we would have to guess what they might correspond to. Talking about the 'genetic code' is also an example of hidden teleology by giving DNA sequences a significance in meaning that they don't have as an intrinsic property.

Reasons. In Chapter 9 I will explain the role of reasons and contextual logic in causation by the environmental context in which an organism functions. Here, I will simply note its existence.

There are therefore multiple categories of causation. In any particular biological function there may also be many separate causes in each category. Complex biological functions have correspondingly complex causal logic. Be wary therefore when someone tells you that a particular fraction of a function or disease is due to a particular cause, such as gene mutations or the environment. The fractions of causation are dependent on each other. If any cause is a necessary cause, the fractions of causation by all the others will fall to zero if that cause is absent. It is a common mistake in reductionist accounts to attribute particular percentages to genetic and environmental factors.[24] When the processes depend on the interactions between different causes and their consequences, the outcome will not be a simple linear sum of the correlations; it will be difficult to predict and can even be counterintuitive.

Conclusions

In this chapter we have seen that the principle of Biological Relativity is best stated as the theory that all levels in organisms have causal efficacy. There is no privileged level from which all the others may be derived. The principle does not, however, mean that all levels are equivalent. The nature of the causal influence of each level on the others may differ.

The principle is a necessary step in understanding living organisms since they are multi-level, open stochastic systems in which the behaviour at any level depends on causal interaction between higher and lower levels and cannot be understood in isolation. That is true even of unicellular organisms and it is even more so of multicellular forms of life, including ourselves as humans.

In this chapter we have seen that incorporation of this principle is necessary even for reductionist explanations to work. The reason is that although the mechanisms by which the components of a system operate might be described by determinate mechanisms, the constraints will appear in the initial and boundary conditions needed to solve the equations for the behaviour of the whole system. Even a nineteenth-century determinist following the ideas of Laplace would have to acknowledge that fact. It is represented in the equations of calculus developed by Newton, Leibniz and Laplace.

When we add the roles of stochasticity and the emergent forms of natural purposiveness at different levels of organisation we arrive at the full principle of Biological Relativity.

The principle can be seen to change fundamentally our view of living organisms. The consequences for biological explanation and for theories of evolution will be explored in the next two chapters.

Those consequences are to be found at all levels in organisms. Although in this book I often focus on the need to revise the gene-centric view to take account of causation at all levels, the same point applies at all levels. Organs are subject to the whole-body systems; tissues and cells are subject to the organs of which they are parts; molecules are subject to all the higher levels. And all of these are subject to the sociology of interactions of organisms with each other and with different species. That is what life is about.

Notes

1 The Latin text is 'nec scire posset, quomodo partes omnes ab universali natura sanguinis moderantur, et invicem, prout universalis natura sanguinis exigit'.

2 In chapter 6 of *The Origin of Species* (entitled 'Difficulties of the theory').

3 Some critics of my recent work towards writing this book have even challenged my right to do so. To quote one of them: 'Has Noble ever produced a single valid criticism of modern evolutionary theory?' I have been doing so for over 40 years. I organised the first debate in Oxford on *The Selfish Gene* when it appeared in 1976. At the International Congress of Physiological Sciences I co-edited the 1993 *The Logic of Life* (Oxford University Press, Oxford) and co-authored a chapter entitled 'The challenge of integrative physiology'. The Nobel laureate Sir James Black and the evolutionary biologist Stephen J. Gould also made important contributions to that book. Many of the ideas of the present book were first outlined a decade ago in Noble, D. (2006) *The Music of Life* (Oxford University Press, Oxford). Since then I have published over 20 substantial peer-reviewed papers on the subject. These are listed in the notes to the Preface of this book and they contain many criticisms of the dogmatic forms of Neo-Darwinism. My main original contribution to the field is to have shown mathematically how strong forms of genetic buffering occur. Models of the cardiac pacemaker can be used in reverse-engineering mode and they reveal how misleading knockout experiments can be in relating genomes to phenotypes. Noble, D., J.C. Denyer, H.F. Brown and D. DiFrancesco (1992) Reciprocal role of the inward currents ib, Na and if in controlling and stabilizing pacemaker frequency of rabbit sino-atrial node cells. *Proceedings of the Royal Society B* 250:199–207; Noble, D. (2011) Differential and integral views of genetics in computational systems biology. *Interface Focus* 1:7–15. This work forms a fundamental challenge to the way in which most studies of gene function have

been performed and therefore to the very basis of gene-centred theories of biology.

4 People still spoke in awe of Bayliss and Starling who had discovered the first hormone, secretin. A.V. Hill (Nobel laureate 1922) was still working in the Physiology Department while I was a student. Henry Dale (Nobel laureate with Otto Loewi in 1936) was sometimes to be seen. Bernard Katz was doing the work that gained him the Nobel Prize in 1970. Peter Medawar was doing what gained him the same accolade in 1960. James Black, who later became Head of Pharmacology, was awarded the Nobel Prize in 1988. Hugh Huxley was doing his pioneering work on muscle contraction. Andrew Huxley arrived as Head of Department in 1960 and received the Nobel Prize in 1963. It is easy to see why I hesitated about moving to Oxford in 1963.

5 This story has been fictionalised in a short story by the award-winning novelist Alison MacLeod: www.amazon.co.uk/ Heart-Denis-Noble-Comma-Singles-ebook/dp/B00GT2G5I4.

6 The reader should understand that I am not proposing that mathematics, in itself, is reductionist. In the beginning of this chapter I am describing what I *thought* (incorrectly!) as a student in the 1960s. As the story develops it becomes clear that mathematics was actually my way out of the reductionist mind-set. Mathematical thinking has a very strong history of addressing complexity and emergence. See, for example, the special issue of *Progress in Biophysics and Molecular Biology* devoted to the topic of Biomathics (volume 119, 2015).

7 Hampshire, S. (1951) *Spinoza* (Pelican Books, London).

8 The French text reads 'Si on connoissoit quelles sont toutes les parties de la semence de quelque espece d'Animal en particulier, par exemple de l'homme, on pourroit déduire de la seul, par des raisons entierement Mathematiques et certaines, toute la figure & conformation de ses membres' (*de la formation du fœtus*, para. LXVI p. 146; https://archive.org/stream/lhommeetlaformato00desc#page/ 146/mode/2up).

9 Descartes was not entirely consistent in his reductionist view of organisms. See Hutchins, B.R. (2015) Descartes, corpuscles and reductionism: mechanism and systems in Descartes' physiology. *The Philosophical Quarterly* 65:669–689.

10 Waddington, C.H. (1957) *The Strategy of the Genes. A Discussion of Some Aspects of Theoretical Biology: With an Appendix by H. Kacser* (George Allen and Unwin, London).

11 See the articles resulting from a Royal Society meeting on top-down causation: Ellis G.F.R, D. Noble and T. O'Connor (2012) Introduction: top-down causation – an integrating theme within and across the sciences? *Interface Focus* 2:1–3.

12 The different forms of causation are classified and dealt with later in this chapter. By active causation I mean efficient cause. The difference between passive and active causation is rather like the difference between anatomy and physiology. Anatomical structure is what permits a function to be performed, which is a formal cause. Physiological processes are what makes it happen. Molecular sequences in nucleic acids and proteins are forms of molecular anatomy.

13 This diagram is highly simplified to represent what we actually solve mathematically. In reality, boundary conditions are also involved in determining initial conditions and the output parameters can also influence the boundary conditions, while they in turn are also the initial conditions for a further period of integration of the equations. There are also important differences between ordinary differential equation models and partial differential equation models. The boundary conditions in partial differential equations become incorporated into the parameters in equivalent ordinary differential equation models. Noble, D. (2012) A theory of biological relativity: no privileged level of causation. *Interface Focus* 2:55–64.

14 I have deliberately included Rosalind Franklin. See http://en .wikipedia.org/wiki/The_Double_Helix.

15 Schrödinger, E. (1944) *What is Life?* (Cambridge University Press, Cambridge).

16 The protein measured is a stem cell agonist called sca-1.

17 Chang, H.H., M. Hemberg, M. Barahona, D.E. Ingber and S. Huang (2008) Transcriptome-wide noise controls lineage choice in mammalian progenitor cells. *Nature* 453:544–548.

18 Williams, R.J. (1956) *Biochemical Individuality* (University of Texas Press, Austin, TX).

19 The term 'conditioned arising' refers to processes that happen in dependence on multiple causes that, to use the language I favour,

form the initial and boundary conditions for each stage of the process. Together with stochasticity in the nature of those causes and the problem of combinatorial explosion in multiple causation, such processes can easily become unique in each instantiation. Re-running the process would not necessarily give the same result unless there is a powerful attractor that makes the final result relatively independent of the initial causes.

20 An important theoretical question is whether this stochasticity is real or apparent, particularly since it is possible to show that determinate input to mathematical formulae called delayed differential equations can produce stochastic results: Lei, J. and M.C. Mackey (2011) Deterministic Brownian motion generated from differential delay equations. *Physical Review E* 84:041105. At the levels of cells, tissues and organisms, I doubt whether this matters.

21 In Aristotle's *Physics* II, 3, and *Metaphysics* V, 2.

22 There sometimes seems to be an irreducible fuzziness to the function–purpose distinction. Both are emergent. But the richness of the emergent behaviour is so different at the different levels that it is possible to draw a line between them, at least in some well worked out cases. The example of cardiac rhythm that I give here in the main text is useful because the function (rhythm) and purpose (to pump blood) emerge at clearly different levels of organisation. 'Function' and 'purpose' are used a lot in physiology. Is this a matter of explanation or of utility, or perhaps both? In a very well thought-through article on the concept of function in modern physiology, Etienne Roux concludes: 'the main interest of functional analysis does not reside in its explanatory but in its heuristic value, opposing the poverty of functionalism as an explicative doctrine to its fruitfulness as a programme for research (Hempel, 1965). In this view, postulating the teleological dimension of the system under investigation is fruitful because it legitimates a reverse engineering-based methodology of research.' Roux, Etienne (2014) The concept of function in modern physiology. *Journal of Physiology* 592:2245–2249. I agree, and the examples from work on the heart that I give in this book illustrate that perfectly. Without reverse engineering, the identification of ivabradine as a useful medication would have been much harder (see Chapter 8 for this example).

23 There is a whole field of mathematics concerned with attractors. There is a useful introduction on Wikipedia: http://en.wikipedia.org/ wiki/Attractor.

24 Remember the equations DNA = 0; environment = 0; DNA + environment = 100%. These express the fact that DNA alone does nothing to produce life. So does the environment in the absence of DNA. The two together ensure that there is life. You might think that scientists are the last people to be confused about the logic of causation. Sadly, that is not the case. Consider a recent article on the causation of cancer published in the prestigious journal *Science*: Tomasetti, C. and B. Vogelstein (2014) Variation in cancer risk among tissues can be explained by the number of cell divisions. *Science* 347:78–81. The authors' experimental observations are fascinating. They show a strong correlation between cell division rate and chances of cancer development, which is already well-known. But they conclude 'These results suggest that only a third of the variation in cancer risk among tissues is attributable to environmental factors or inherited predispositions.' This is strange since we know, for example, that smoking is a major cause of cancer, with over 80% correlation; much more than one-third in the case of lung cancer. The problem is solved once one realises that the relevant logic is: if X and if Y then Z. In this logic if either X or Y do not occur then Z does not occur. You would therefore find 100% correlation with either X or Y. But to conclude that the other contributes 0% would obviously be wrong. The same kind of combinatorial logic applies when the correlations are less than 100%. In combinatorial logic, percentages do not add linearly.

7

Dancing Nucleotides
Natural Genetic Engineering

The genome is an organ of the cell.
Barbara McClintock, Nobel Prize lecture 1983

Pipes and Templates

For more than a decade now I have been privileged to learn from and sometimes even perform with the Swiss virtuoso classical guitarist, Christoph Denoth. The title of one of my previous books, *The Music of Life*, was chosen during a visit to Basel in 2003. Christoph was in the audience for my lecture and I wanted a metaphor that he would appreciate. Music as a metaphor for the processes of life emerged when, on the spur of the moment, I compared the genome to a vast pipe organ. The point is that the pipes do not themselves play the music. They are passive agents. The organist is the active agent. As we have learnt in earlier chapters, DNA sequences are also passive. It is the organism that actively 'plays' them.

A few years later, Christoph and I were at the workshop of Paul Fischer, in the charming Oxfordshire town of Chipping Norton. Paul was making a new guitar and we were there to assist in choosing the wood. As we tapped the sheets he brought out from his store of well-seasoned spruce we listened to the resonance of the pieces that might be used for the all-important soundboard. This is one of the most critical steps,

Figure 7.1 Paul Fischer guitar (above) and the template used to construct it
(bottom). The lines forming a fan indicate where the struts are placed
underneath the soundboard to influence the resonance of the instrument
(courtesy of Paul Fischer, luthier).

determining how the finished guitar will sound. There are people who
specialise in finding the Swiss Alpine trees from which to cut the pieces
with the required fine and uniform grain that top luthiers require.

A few months later the guitar was ready to be played. Given all the
care and attention in the hands of a master craftsman it was not surprising that the sound was a joy to listen to. All that remained were a few
fine adjustments and the instrument would be ready for concert perfor-
mances. Except that when on delivery Christoph played it the sound was
muffled, dead. This was very strange. We had ourselves witnessed the
sweetness of the sound and felt the joy of playing it in the workshop.[1]

We telephoned Paul, who was equally surprised, until he asked us
a question. Can you feel inside under the soundboard? There was the
explanation: a wooden spacer to hold the struts during construction had
accidently been left in place. On removing it, the sweet sound reappeared.
I kept the spacer but the next time I phoned him to return it he told me
the sad news that this was to be one of the last guitars he would make. He
had decided to retire, and he offered to let me have the template he had
used for so many of his guitars. It is a simple plastic shape to determine
the shape of the soundboard. It also has lines that determine where those
wooden struts will be placed in a fan arrangement (Figure 7.1).

The precise placing of these fan struts (braces) determines the balance between the resonances of the wood at different frequencies. In a sense, therefore, the template is a cause of the way in which the instrument plays. So also is the way in which the guitarist strikes the strings. Even the shape, length and smoothness of the fingernails are important. But clearly we are dealing here with different kinds of cause. The template is a passive cause. Its existence and use were necessary but in no sense can the template be said to play the music. Just as with the keyboard of a pipe organ, the active cause of the music that the instrument produces is the act of playing by the performer. This distinction between active and passive causes is important.

Genome sequences work just like Fischer's template. They are necessary; no doubt about their importance. As James Watson first suggested, they form a sequence template that is used to enable protein sequences to be formed.[2] They are for that reason totally necessary for life as we know it to exist since life on Earth requires proteins. But the DNA sequences do not themselves carry out the physiological functions served by the active networks of proteins, metabolites and the three-dimensional cellular and organ structures within which they work.

It's as simple as that. With that distinction in mind, let's now see how we can develop a theory of evolution that respects it.

Summary of the Problem

In Chapter 5 we found that Neo-Darwinism is incomplete as a theory of evolution. It also suffers from deep conceptual confusions, and is not compatible with the wider range of experimental evidence we now have. To summarise the conceptual confusions:

First, it is seriously confused about the central entity of the theory, the gene. Different definitions of a gene and of genetics have incompatible consequences for the theory since the definition of a gene has shifted all the way from referring to a necessary cause of a particular phenotype to referring to a DNA sequence, where the issue of causality is an empirical question.

Second, it is confused about the different forms, natures and direction of causality, which in turn is why it prefers to avoid or 'reduce' the role of teleology and purpose. This is also why it represents genes (DNA sequences) as active causes rather than as passive templates. Significantly, the teleological language ('selfish' etc.) has not been eliminated; it has simply been wrongly attributed to the gene instead of to the organism. Of course it is true that most active scientists who think in Neo-Darwinist terms would deny that the anthropomorphic language is intended; they would say it is simply a way of speaking. But that is not entirely true. Representing DNA sequences as active causes 'determining' the organism is central to the language of Neo-Darwinism. The active nature of this representation is clearly intended.[3] It is a hidden form of anthropomorphism and is one of the reasons for the powerful influence Neo-Darwinist metaphors have had on social theory, for example. The implication is that what is thought to apply to genes justifies similar but often more literal language in social science. Even if such language were justified at the level of genes it doesn't follow that it is justified at any other level.

Yet, as we will see in this chapter and the next one, teleology is alive and well because there is what we may call natural purposiveness,[4] but it belongs to the organism not its DNA. Representing DNA sequences as passive templates would avoid the misleadingly anthropomorphic language of the gene-centric popularisations.

Finally, Neo-Darwinism uses an outdated concept of programming to defend the concept of a genetic programme. There are no complete 'programmes' within the genome. All the logic of life necessarily includes other components of organisms.

These are confusions of the kind that reveal when it is necessary for a complete rethink. That is also the purpose of this chapter. The essential points can be summarised as follows:

1. Recognise that the Weismann Barrier is a relative barrier rather than an absolute one.
2. Recognise that genetic variation is not always random with respect to function.
3. Recognise the existence of other forms of inheritance in addition to strict Mendelian inheritance.

4. Recognise that the Central Dogma of molecular biology is better represented as an important chemical fact about coding, rather than an absolute statement about control by and primacy of the genome.

5. Recognise the full significance of mobile genetic elements and the reorganisation of genomes.

6. Recognise the inheritance of epigenetic and similar Lamarckian forms.

7. Recognise the significance of symbiogenesis and many other forms of co-operation.

8. Recognise the significance of niche construction and the active role of organisms in evolution.

9. Recognise that evolution is a multi-mechanism process, that the Neo-Darwinian mechanism is just one of them, and that we really do not yet know the relative contribution of each process to each stage of evolution. This would be a return to Darwin's more nuanced view that other processes may also exist.

10. Respect the principle of Biological Relativity. The relativity principle applied to evolution is that nature may use mechanisms and selection at any scale, not just molecular. There is no privileged mechanism or scale for the transformation of species.

It is important to note that all of these criteria depend on experimental evidence. This chapter will give an account of experimental evidence that challenges the Weismann Barrier and the Central Dogma. Chapter 8 will then do the same for epigenetics and related processes. These two chapters therefore complement the conceptual analysis of Chapter 5 by using the principle of Biological Relativity outlined in Chapter 6.

The Weismann Barrier is Relative, not Absolute

In Chapter 5 I described Weismann's fundamental contributions to the two key assumptions of Neo-Darwinism: random causes of variation and isolation of the germ line, which is the Weismann Barrier. I also gave a partial answer to the question of how those ideas could have become so strongly held on the basis of experimental evidence that could not

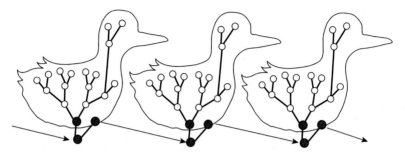

Figure 7.2 Representation of the Weismann Barrier. The black circles represent the germ line cells. The white circles represent the soma cells. Germ line cells are used to create soma cells, but soma cells are never allowed to influence the germ line cells. This is the essence of Weismann's germ-plasm theory. From: Hardy, Sir Alister. *The Living Stream*. Collins, London, 1965 (p. 76).

even satisfy the basic criterion of being a valid test of the main idea of Lamarckism. The answer was that the evidence from Weismann's tail experiments was never thought by him to be sufficient and that he had formulated the essential principles of Neo-Darwinism well before he performed those experiments (Figure 7.2).

We have to look elsewhere for the origin of Weismann's ideas. To do this I will draw on some of the conclusions of Chapter 1. Neo-Darwinism is a product of nineteenth-century scientific thought in the last decades before the revolutions in physics leading to quantum mechanics and relativity theory. As we saw in Chapter 6, this period was the heyday of mechanistic thought. From this perspective the universe is a giant piece of clockwork. It was seen as hard, irreducible elements, the postulated indivisible atoms, bouncing into each other like billiard balls, and influencing their movements also by forces at a distance, like gravity and electromagnetism, to produce a process that, in principle, could be completely predicted from classical mechanics. This is the origin of the Laplacian 'god's-eye' view. An all-knowing intelligent being could, through the equations, understand precisely what would happen in the future, and what must have happened in the past. All that was required was the data on the positions and velocities of all the elements. As we saw in the last chapter, that is both the loophole in the theory and, even more serious, it is an impossible condition.

But, for the moment, let's ignore those problems and explore this viewpoint because it was so profoundly influential that even biologists who appreciated the systems approach were captivated by its system of thought. To illustrate this influence particularly as it affected biological thought, consider the work of the greatest physiologist of the nineteenth century, Claude Bernard.

Claude Bernard was a consummate experimentalist. He revealed the functions of the liver, pancreas and other organs of the body. He pioneered the understanding of digestion. In many ways he can be said to have initiated the field of experimental physiology, and was widely appreciated to have done so when the Physiological Society in the UK was founded in 1876. He was also a great theoretician. His book, *Introduction to the Study of Experimental Medicine*, published in French in 1865, was ground-breaking in the sweep of its ideas. Well before the great systems physiologists of the twentieth century, he formulated the principle of homeostasis, according to which living organisms are characterised by the control they exert over their bodies. He called this the maintenance of the constancy of the internal environment. By internal environment he meant the liquid environment of the cells, tissue and organs of the body. Temperature (in warm-blooded animals), pressure and many other variables, now known to include acidity, salt concentrations and metabolite concentrations, are all kept within certain bounds by feedback processes. Although he did not use the term feedback, this is what his theory entailed. I believe therefore that he has a good claim to be the first systems biologist in the modern sense of the word.[5]

He even foresaw that this would require the use of mathematics in biology: '[The] application of mathematics to natural phenomena is the aim of all science, because the expression of the laws of phenomena should always be mathematical.' This is the quotation I used in the brief autobiographical sketch at the beginning of the last chapter in the context of the ultimate form of reductionism. I repeat it here in a different context: that of the pervasive influence of the mathematically precise clockwork view of the universe in the mid-nineteenth century, described in Chapter 1.

The influence of Laplacian determinism can also be seen in his arguments against those who he suspected of reviving the idea of vitalism. Bernard was insistent: to suppose that the world was anything other than mechanically determinate would be to deny the very possibility of science. We can now see that this is far too restrictive as a view on what can count as science.[6] After the revolutions in physics produced by relativity theory and quantum mechanics, science managed perfectly well with probabilities rather than certainties. Given the extent of the stochasticity we have discovered in biology, it too has to deal in probabilities. A nineteenth-century systems physiologist like Claude Bernard has therefore to be seen in the context of his period.

So should August Weismann. I believe that this is the important clue to some of Weismann's ideas, which also need to be seen in the context of the period in which he lived. Weismann was born about two decades after Bernard, but he also must have been greatly influenced by nineteenth-century certainties amongst scientists concerning the determinate materialistic nature of the universe. As Weismann's tail-amputating experiments showed, he was also determined to oppose what he regarded as occult, even mystical ideas, including the weirder forms of Lamarckism.

This motivation carried right through to the formulation of the Modern Synthesis, and to the dogmatic forms of Neo-Darwinism that we have seen today in its popularisation. The revolution in physics passed by unnoticed. An important message of this book is that it should not have done so. Applying the relativity principle to biology produces a complete change in viewpoint. As we will see in this and the next chapter, the genome is not completely isolated. The Weismann Barrier should be seen as a relative – not absolute – barrier in the minority of organisms with a separate line of germ cells. In these organisms special mechanisms are required to circumvent the barrier. Such mechanisms are now known to exist. Weismann was not wrong to identify the barrier. The mistake was to treat it as absolute.

Genetic Variation is not Random

Defining what is meant by 'random' is itself a major field of enquiry in mathematics, computation and in science generally. The question of whether there are truly random events in the universe is a vexed one. As we saw in Chapter 1, this question lies at the heart of theories of quantum mechanics. Probably, we will never know, perhaps cannot know in principle, the answer to that kind of question. But it warns us that defining randomness is not easy.

The best way to sidestep these deeper problems is to ask an easier and more specific question: 'random with respect to what?'. In evolutionary theory that makes the problem much simpler. Both Neo-Darwinists and their opponents can then agree that what is really meant is 'random with respect to physiological (phenotypic) function'. That is so because one of the central tenets of Neo-Darwinism is the exclusion of any form of Lamarckism. By contrast, a Lamarckian must maintain that at least some changes are not completely random with respect to function.

We can approach this question in three stages. The first stage is to establish that genomic change is not random with respect to location in the genome. The reason for asking that question first is that without establishing that there are preferred locations of change, the argument for any kind of functionally relevant change in the genome cannot even get off the ground. The only way in which such a change can occur is through influencing the physical and chemical properties of the genetic material. Preferred locations of change are therefore a pre-condition for functional change in genome sequences to be possible. If all locations in the genome were equally open to changes, there would be no possibility for functionally relevant change.

Of course, demonstrating the existence of hotspots and other ways in which change is not randomly distributed with respect to location does not, in and of itself, demonstrate any form of functionally relevant change. The existence of hotspots could be simply a consequence of the physico-chemical properties of the genome and its associated proteins even if no functionally relevant changes occur. Further experimentation is required to determine whether evolution could use such mechanisms in a functional way.

The second stage in the argument is to note that well-documented examples of functionally relevant genomic change already exist. The best-investigated case is the evolution of lymphocytes. The germ line has only a finite amount of DNA. In order to react to many different antigens, lymphocytes 'evolve' quickly to generate extensive antigen-binding variability. There can be as many as 10^{12} different antibody specificities in the mammalian immune system, and the detailed mechanisms for achieving this have been known for many years. The mechanism is directed, because the binding of the antigen to the antibody itself activates the proliferation process. The antigen activates special lymphocytes (cells in the blood stream) called B-cells, which evolve rapidly to generate a huge range of antigen-binding variability.[7] Targeted speeding-up of change is therefore one mechanism by which functional change can occur. That is true even if the individual changes at that location are random. The functionality lies in the targeting of the location. That targeting is not random.

A possible objection to this example would be to say that the non-random occurrence of hotspots in the genome was itself a result of evolution, which could have been the result of random change followed by natural selection. Indeed, that is possible. But it remains true to say that the organism is thereby endowed with a natural purposiveness, however that may have arisen. The organism can use that natural purposiveness to seek non-random changes in response to an environmental stimulus. There is a kind of bootstrapping here. Once a form of natural purposiveness has evolved it can be used to ensure that subsequent evolutionary change is not generated randomly.

The example from the immune system is not, of course, an example of trans-generational inheritance. The immune system evolves in the lifetime of an individual and, in the case of mammals, the period spent in the womb.[8] We will now consider examples that resemble the evolution of the immune system, but also involve trans-generational effects.

A well-known functionally driven form of genome change is the response to starvation in bacteria. Starvation can increase the targeted reorganisations of the genome by five orders of magnitude, i.e. by a factor of over 100,000. This is one of the mechanisms by which bacteria can evolve very rapidly and in a functional way in response to environmental stress. As we saw in Chapter 4, some forms of DNA transfer in microorganisms are also functional since they are activated as a response to a stressful environment. That is precisely when organisms need to find a new solution to the problem they have encountered.[9]

A similar targeting of location where genomic change can occur has been found in experiments on genetically modified fruit flies. One of the common ways in which genetic modification is achieved is to use a particular kind of mobile genetic element that can move around the genome using a cut-and-paste mechanism that does not require an RNA intermediate. Most often the insertions occur in a random way. But when DNA sequences from certain regulatory regions are used, they get inserted preferentially near the gene from which the sequence was derived.[10] This process targets the changes in a way that is clearly not random with respect to possible function.

The third stage in the argument is experimental demonstration that the inheritance of acquired characteristics occurs. There are now many examples of that, some of which I will use as examples in a later section and in the next chapter. Those experimental results require functional inherited change, either genetically or epigenetically, even if we do not yet know the molecular mechanisms.

All these stages of the argument are necessary.

Misinterpretations of the Central Dogma

The Central Dogma of molecular biology was originally formulated by Francis Crick in 1956 following the discovery of the triplet code, in which each amino acid in a protein corresponds to three bases in the DNA or RNA. The coding between nucleic acid sequences and amino acid sequences works only one way. Triplets of nucleic acids in DNA or RNA code for an amino acid, but amino acid sequences in proteins do not code for nucleic acid sequences or for amino acid sequences in other proteins. The discovery of the triplet code was a major step forward. But the choice of the word 'dogma' was unfortunate.[11] Science does not deal in dogmas. The use of the word in this particular case has been particularly confusing because the experimental observation has also been widely misinterpreted.

It seemed to many people that the Central Dogma confirms the Weismann Barrier idea. But this is not and cannot be the case. The dogma was misinterpreted to mean that information could not pass from the organism and environment to the genome. To quote *The Selfish Gene*, genes are 'sealed off from the outside world'. This is simply incorrect, and it is not even justified by the Central Dogma. It is important therefore to note that Crick's original statement of the dogma was qualified in two very important respects:

> The central dogma of molecular biology deals with the detailed residue-by-residue transfer of sequential information. It states that *such information* cannot be transferred back *from protein* to either protein or nucleic acid.

I have italicised '*such information*' and '*from protein*' since it is evident that the statement does not say that *no* information can pass from the *organism* to the genome. Crick would have known that absolute isolation of the genome from control information could not be true. How else could the same genome be used by the many different cells, tissue and organs of the body to generate very different phenotypes? Note also that the statement refers to transfer back *from proteins*. The information that regulates gene expression via transcription factors and epigenetic marks comes, of course, from the networks *as a whole*, not from individual proteins, although the final message is conveyed by proteins called transcription factors. It is the *pattern* of such factors that is important, and that it is a global property of the cells, tissues and organs involved. There are many possible patterns of transcription factors, each of which corresponds to a different phenotype outcome. The information that passes from the system to the genome is of a different kind to that involved in coding. It is not a property of individual molecular sequences, but rather a property of an ensemble.

In 1956 it must have seemed secure at least to interpret the Central Dogma to exclude changes in the DNA code itself. The control might have been restricted to determining gene expression patterns, so leaving the sequence itself secure against changes. The discovery of the process of transcription of RNAs back into DNA (using an enzyme appropriately called reverse transcriptase) shook that assumption and led Crick to modify the statement in 1970 to include this possibility. It seemed, however, that the use of the Central Dogma to support the Weismann Barrier might still be secure since the reverse transcription was from RNA sequences, not from protein sequences. The revision of the dogma is illustrated in Figure 7.3.

Notice, however, that the diagram is incomplete since it does not include any flow of information from the organism that controls patterns of gene expression, and which might initiate or control insertion of DNA into the genome either directly or following reverse transcription from RNA. Yet, by reverse transcription, it becomes possible to transfer sequences, including whole domains corresponding to functional parts of proteins, from one part of the genome to another. As we will see in the next section, that must have happened during evolution. The idea that the genome is isolated from any functional influences on the sequences is therefore simply incorrect.

I sometimes describe this misinterpretation as the great mistake of molecular biology. There is nothing wrong with the molecular biology experiments and the observations on the direction of coding. That is a straightforward and correct chemical observation. The mistake was one of interpretation, that the results confirm the Weismann Barrier

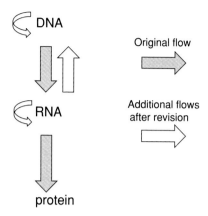

Figure 7.3 Central Dogma after discovery of reverse transcription. DNA codes for RNA which then codes for proteins (white curved arrows). RNA can be reverse transcribed into DNA (upward white arrow). This kind of diagram is often described as defining the information flows in biological systems. But it omits information flows that control gene expression, i.e. transcription factors, methylation and interactions with histones. It also omits the use of the molecular mechanism of reverse transcription to reorganise the genome. The circular arrows represent the fact that DNA can also be involved in cut-and-paste modifications of the genome without the involvement of RNA, and that a similar self-templating can occur in RNA.

assumption in Neo-Darwinism. The discovery of reverse transcription should have been a warning signal. That reverse transcription has immense significance for evolutionary biology was soon noticed by the greatest logician of science of the twentieth century: Karl Popper.

It is not widely known that Popper gave an important lecture to the Royal Society in 1986, entitled 'A new interpretation of Darwinism'. It was given in the presence of the Nobel laureates Sir Peter Medawar, Max Perutz and other key figures, and it must have shocked his audience. He proposed a completely radical interpretation of Neo-Darwinism, essentially rejecting the Modern Synthesis by proposing that organisms themselves are the source of the creative processes in evolution, not random mutations in DNA. He said that Darwinism (but I am sure he meant Neo-Darwinism) was not so much wrong as seriously incomplete. He saw that reverse transcription, which is the process that allows DNA segments to be transported from one region of the genome to another via an RNA intermediate, greatly weakens the Central Dogma. In particular, it weakens the dogma in justifying Neo-Darwinist theory since it changes the genome from the read-only idea of Neo-Darwinism to a read–write genome.

He was therefore deeply suspicious of sophisticated manoeuvrings and redefinitions to protect the dogma from falsification. In his 'conjectures and refutations' view of science it is better to acknowledge when a strong version of a theory has been refuted. The strong Neo-Darwinist interpretation of the Central Dogma was refuted. But he went further than this. He saw that reverse transcription could be one of the routes through which Lamarckian processes and wholesale reorganisation of genomes could occur. Again, the philosopher in him wanted to see this recognised, not hidden behind a web of clever re-interpretations. I will return later in this chapter to the reasons why Popper's lecture is not widely known.[12]

Mobile Genetic Elements

Barbara McClintock was an American plant biologist working on maize (Indian corn) in the 1920s and 1930s. She was visualising chromosomes when she noticed that chromosomes recombine and exchange information during reproduction. In the 1940s and 1950s she made the groundbreaking discovery of transposition of genes. Parts of the genetic material could move from one location to another. Her work was received with such deep scepticism that she ceased publishing on her major discovery in 1953. Thirty years later, at the age of 81, she was awarded the Nobel Prize for Physiology and Medicine for the discovery of mobile genetic elements.

In an article in *Science* based on her Nobel Prize lecture she wrote:

> In the future attention undoubtedly will be centered on the genome, and with greater appreciation of its significance as a *highly sensitive organ of the cell*, monitoring genomic activities and correcting common errors, sensing the unusual and unexpected events, and responding to them, often by *restructuring the genome*. We know about the components of genomes that could be made available for such restructuring. We know nothing, however, about how the cell senses danger and instigates responses to it that often are truly remarkable.[13]

The italics are mine. Her statement that the genome is a highly sensitive organ of the cell gets causation the right way round. It is cells, and organisms, that display active behaviour. The genome is passive until activated. She also realised the importance of her discovery of mobile genetic elements in the ability of organisms to restructure the genome.[14]

The mechanisms of transposable elements therefore illustrate one of the important breaks with the Central Dogma of molecular biology. Retrotransposons are DNA sequences that are first copied as RNA sequences, which are then inserted back into a different part of the genome using reverse transcriptase. DNA transposons may also use a cut-and-paste mechanism that does not require an RNA intermediate.

We now know that these mechanisms are used extensively in genome evolution. In a major book on *The Concept of the Gene in Development and Evolution* in 2008, Peter Beurton and his co-authors write: 'it seems that a cell's enzymes are capable of actively manipulating DNA to do this or that. A genome consists largely of semi-stable genetic elements that may be rearranged or even moved around in the genome thus modifying the information content of DNA.' The Australian specialist on RNAs and plasticity John Mattick expressed a similar sentiment when he wrote 'the belief that the soma and germ line do not communicate is patently incorrect.'[15]

The Central Dogma, as a general principle of biology, has therefore been progressively undermined until it has become useless as support for the Modern Synthesis or even as an accurate description of what happens in cells. All that remains is the chemical fact that protein sequences do not code for nucleic acid sequences, which is what the impressive molecular biological experiments actually showed. That chemical fact, together with the triplet coding, and the roles of RNAs as intermediates, all remain as great achievements, but they do not justify a DNA-centric view of life.

Natural Genetic Engineering: Genome Reorganisation

The experimental evidence that wholesale reorganisation of genomes has occurred during evolution shows that genomes did not always evolve by gradual accumulation of random small mutations. Some of the best evidence has come from genome sequencing of different species. Of course,

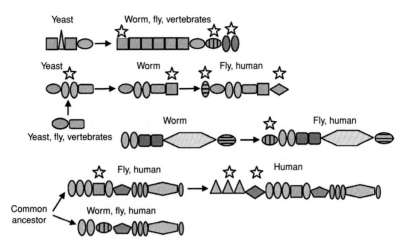

Figure 7.4 Evidence for how gene domains must have moved around during evolution to form different functional proteins by domain accretion. This figure shows diagrammatic representations of three groups of individual proteins that contribute to the formation of chromatin, the backbone of chromosomes. Each coloured shape shows a single protein domain. Each protein is made up of several domains. In each case, as we move from yeast (unicellular) to worm, fly, various vertebrates and the human, the number of domains increases. The significant fact is that the development from one to the other did not occur by gradual accumulation of small mutations. Instead, whole domains hundreds of amino acids in length have come together to form the new protein. The red stars show the domains that must have moved in this way. It is extremely unlikely that this result could have been achieved by gradual accumulation of small random mutations. Reproduced with permission. For full details of the domains and proteins involved, see the original *Nature* paper at www.nature.com/nature/journal/v409/n6822/fig_tab/ 409860a0_F42.html. For a colour version of this figure, please see the plate section.

the species that exist today are not the historical species that formed the branching 'tree of life' or the modern network version (Figure 4.10) during evolution, but their genomes nevertheless contain the tell-tale signs of what must have happened. Figure 7.4 is based on an illustration in the 2001 *Nature* article reporting the first full draft sequence of the human genome,[16] and it compares the structure of certain proteins in a variety of species, including yeast, worm, fly, vertebrates and human.

The two classes of proteins that show this phenomenon of domain accretion very clearly are chromatin binding proteins and transcription factors, both of which are involved in the structure and regulation of the genome. The structure of the chromatin determines how loosely or tightly the DNA is wound around it. In turn, that determines how easily the proteins that form the transcription apparatus can move along the DNA molecular thread to read it. Significantly, organisms have reorganised their genomes through natural genetic engineering, forming templates for the proteins that are most intimately involved in controlling the genome and the way in which it is read.

The standard Neo-Darwinist interpretation of this discovery is to call such reorganisations of the genome 'large mutations'. That obscures their great significance. The domains have functionality. They might be essential for a particular protein receptor, or its overall structure, possible conformation changes or any other functional property. These 'large mutations' are therefore variations that move and create new proteins that are much more likely to be functional than creation via accumulation of random small mutations. It obscures this fundamental difference to conflate the two mechanisms. Evolution has been quite conservative in producing completely new DNA sequences, which is why so many species share great proportions of their genomes. The differences between a worm, a fly, a mouse or a human have more to do with wholesale reorganisation and regulation of the genome than with completely new sequences. It seems to me to be more honest that we should admit that this is far from what was anticipated by the architects of Neo-Darwinism. In spirit, at least, that theory is a gradualist one so far as mutations are concerned.

We may also speculate that the insertions of whole domains into new parts of the genome may have been far from random with respect to location.[17] A mechanism comparable to that used by the immune system (see pages 195–196) may have greatly improved the chances of new functions arising.[18]

To appreciate the full significance of these mechanisms by which whole domains can be moved around in the genome, imagine two children playing with a construction kit like Lego. To one child we give a pile of the original simple Lego bricks. To the other we also include many preformed shapes. It is obvious that when asked to make any construction that requires the preformed shapes the second child will succeed much faster than the first. In the same way, evolution is much more likely to generate successful novel organisms if it can 'play' with preformed DNA domains. Existing functionality is transferred into forming new combinations.

These arguments and the supporting evidence do not of course exclude the process of evolution of proteins by gradual change. Sequence studies of haemoglobin proteins in different species show that such gradual change also occurs.

Significance of Symbiogenesis and Co-operation

Moving whole functional domains of DNA around in the genome is impressive enough, but what about moving or combining whole genomes? This is the process called symbiogenesis.

The mother of symbiogenesis in evolution was surely Lynn Margulis.[19] I was privileged to interact with her during a year when she came from Amherst, Massachusetts in the USA to be a visiting professor at Oxford University in 2008–2009. I already knew that mitochondria in the cells of animals and plastids in the cells of plants were thought to have arisen from fusion between early forms of unicellular organisms and particular forms of bacteria, the ancestors of mitochondria and plastids. But it was not until that year of intense discussion with Margulis that I realised the full significance of what she and others had done in this important area of biology. She insisted that I should chair what has now become a famous debate between her and Richard Dawkins, recorded by the video team *Voices from Oxford*.[20] It became impossible to think of cells – any cells, in any species – without seeing them as the miracle of co-operation that they represent.

It is hard to think of the evolution of eukaryotes – the kinds of cells that form your and my bodies, packed as they are with nuclei and organelles – as anything other than a major step in evolutionary history. It is what made possible the evolution of highly organised organisms, plants and the whole of the animal kingdom. Yet it is impossible to explain such a critical step as simply the outcome of chance mutations in the genomes of prokaryotes followed by natural selection. Before that evolutionary threshold was passed, it was impossible for organisms with separate germ cell lines to exist. The Weismann Barrier and the original basis of Neo-Darwinist theory would have been irrelevant. Yet the great majority of time during which life has evolved on Earth occurred *before* this threshold. When you realise that, Neo-Darwinism takes on a different historical significance. If and when we do find life on other planets, it is quite possible that Neo-Darwinism would be irrelevant for understanding its evolution. The great majority of life on other planets may never have passed this threshold.

We don't know how rapidly the symbiogenetic events leading to eukaryotes may have happened. A simple stage of ingestion of a bacterium by a prokaryotic cell would have been virtually instantaneous of course: just gobble the goody! But there must have been many other stages.[21] Instead of being eaten up for fuel, the ingested bacterium would have needed to survive inside and prove beneficial to its host, while the host would have needed to be beneficial to the bacterium. We know also that much of the DNA of bacterial origin subsequently moved to the nuclear genome, presumably by the transposition mechanisms discussed earlier in this chapter. Only a small amount of DNA now remains in mitochondria and chloroplasts. That would have involved many steps of natural genetic engineering.[22]

Having absorbed all of these insights, I was naturally intrigued to see what a debate between Dawkins as a prominent Neo-Darwinist and Margulis as a champion of symbiogenesis and other forms of co-operation would reveal. The recording of the debate and transcript[23] are both accessible via the internet, so readers can view and read them to form their own judgements. I think the debate revealed two things of importance in this book. First, a Neo-Darwinist like Dawkins appears initially to see no problem in absorbing symbiogenesis into the theory. The early part of the transcript includes the statement 'Lynn, I don't think you go far enough.' This is a common Neo-Darwinist tactic that seems to explain everything but in so doing explains nothing.[24]

But much later, as I noted in Chapter 5, Dawkins challenges Margulis to justify why she needed to bring symbiogenesis into the picture when Neo-Darwinism is both sufficient and the most parsimonious hypothesis. Of course, it is unfair to pin anyone down to the precise words they may use in an impromptu debate. But, if we look at the way in which Neo-Darwinism developed, and understand the spirit in which it was formulated, then the second intervention is much more in line with that spirit than the initial one. The emphasis in Neo-Darwinism is on chance occurrence and gradual accumulation of small changes. Symbiogenesis, by contrast, is the fusion of whole genomes into a single new organism.[25] It is the very essence of co-operation in nature. The classical Neo-Darwinist cap doesn't fit it. Yet it was responsible for one of the most important stages in the evolution of life on Earth.

I am a biased observer of course, and I was the chairman of the Margulis–Dawkins debate. The conclusions I draw from it are, first, that Richard Dawkins was the more effective debater. But Lynn was surely right to argue that symbiogenesis was responsible for an extremely important step in evolution: the emergence of eukaryotes, organisms with cells like those found in you and me.

Does symbiogenesis still happen, or was it just an early stage in evolutionary history? That depends on what precisely we include in symbiogenesis. It doesn't need to involve the fusion of whole genomes. By this criterion, we humans and all multicellular organisms are symbionts. We co-operate with a vast number of bacteria that serve necessary functions for us, as we do for them. In fact, as I noted in Chapter 4, there are many more bacteria in us than cells with our own genome. Their total DNA complement is at least 250 times larger than our own. Those bacterial populations evolve quickly, which is fortunate if we have to change diet.

Transfer of DNA between bacteria and the host can also occur. It has happened in the case of gut bacteria in insects, one of the best examples of DNA transfer not just between different species, but across different kingdoms of organisms.[26] An even more remarkable case of horizontal DNA transfer involves nucleotide sequences from three separate kingdoms contributing to the mechanisms by which tiny animals called rotifers can resist extreme desiccation. The DNA sequences involved were acquired from fungi and bacteria and they contribute around 8% of the animal's genome.[27] Transfer of DNA between green algae and a sea slug has been shown to be responsible for a remarkable ability of the sea slug to photosynthesise. Algal DNA has not only been transferred to the host, but has been incorporated into the sea slug's genome.[28]

I conclude that symbiogenesis is still alive and well, kicking away successfully, most particularly in our own bellies.[29] It is also responsible for some of the forms of non-Mendelian inheritance.[30]

Moreover, many other forms of co-operation are alive and rampant everywhere in nature and at all levels of biological organisation, a point that has been argued very forcibly in a recent book by Martin Nowak and Roger Highfield.[31]

Conclusions

It is the 'atomistic' gene-centric nature of Neo-Darwinism that is the main problem with the theory. In its modern interpretation, it gives to a molecule, DNA, the primary causal role in evolutionary change, through chance non-functional variations in its sequences. Molecules are important, of course, but they are not alive. Outside an organism DNA can do nothing at all. It is better regarded as forming templates in a database that the organism and its progeny can use. The correct direction of causality is from the organism to its genome. As Barbara McClintock wrote in her 1983 Nobel Prize lecture, 'the genome is an organ of the cell'.

The reason is that the Weismann Barrier is not an absolute barrier. That is so even in the relatively small proportion of organisms that have a separate germ line. The genome is not isolated from the rest of the organism or from the environment. It can be changed by the processes of natural genetic engineering.

This chapter sets the scene for the next one, in which we reveal the extent to which the genome is controlled by the organism through epigenetic processes. It will be through understanding those processes that we can progress to a view of evolutionary biology that respects the principle of Biological Relativity.

Notes

1 The history of this guitar has now come full circle. After playing it for several years, Christoph Denoth generously agreed with me to donate it to my brother, Raymond Noble, who uses it to compose music for medieval Trobador songs for which the music has been lost. He confirms the sweetness of the tone and its bell-like quality. Ray's involvement, as a scientist and philosopher, in the ideas of this book are acknowledged at the beginning of Chapter 6. He is also a brilliant musician as lead singer in the *Oxford Trobadors*.

2 James Watson may have been the first to use the template idea: 'I jokingly used to say, well it was either the template for protein synthesis or it was there to control the viscosity of a cell. Now the

latter was a private joke between Francis and I, 'cause he'd spent several years measuring the viscosity of cells, something which I think he later realized was a total waste of time. So I said it's a template.' Interview at www.dnalc.org/view/15474-RNA-s-role-in-the-cell-James-Watson.html.

3 Richard Dawkins originally thought that 'selfish' was not a metaphor, but to be interpreted literally: Dawkins, R. (1981) In defence of selfish genes. *Philosophy* 56:556–573. The reason given is worth quoting in full: 'that was no metaphor. I believe it is the literal truth, provided certain key words are defined in the particular ways favoured by biologists.' But a metaphor does not cease to be a metaphor simply because one defines a word to mean something other than its normal meaning. Indeed, it is the function of metaphor to do *precisely* this.

4 Some biologists use the term 'natural intelligence' for what I have called 'natural purposiveness'. I have chosen the second form to avoid some of the deeper philosophical questions arising from the use of 'intelligence'.

5 This view of Claude Bernard is developed more fully in Noble, D. (2008) Claude Bernard, the first systems biologist, and the future of physiology. *Experimental Physiology* 93:16–26.

6 'A philosopher once said "It is necessary for the very existence of science that the same conditions always produce the same results". Well, they don't!' Feynman, R. (1967) *Character of Physical Law* (MIT Press, Cambridge, MA; p. 27).

7 For further details, see Shapiro, J.A. (2011) *Evolution: A View From the 21st Century* (FT Press, Upper Saddle River, NJ; pp. 66–68).

8 The period in the womb is when many maternal effects are known to have their influence on the offspring. See Gluckman P. and M. Hanson (2004) *The Fetal Matrix: Evolution, Development and Disease* (Cambridge University Press, Cambridge); Bateson, P., P. Gluckman and M. Hanson (2014) The biology of developmental plasticity and the Predictive Adaptive Response hypothesis. *Journal of Physiology* 592:2357–2368.

9 The mechanisms of increased variation discussed in this chapter form the basis of the theory of facilitated variation proposed by Kirschner, M. and J. Gerhart (2010) Facilitated variation, in

evolution. In *The Extended Synthesis*, M. Pigliucci and G. Müller, editors (MIT Press, Cambridge, MA).

Yoav Soen and his colleagues have proposed that organisms may harness random variation in a process of adaptive improvisation, by which adaptive changes can be transmitted across generations, allowing rapid improvement and assimilation in a few generations. Adaptive improvisation 'provides a basis for Lamarckian adaptation that is not limited to a specific mechanism and readily accounts for the remarkable resistance of tumors to treatment.' Soen, Y., M. Knafo and M. Elgart (2015) A principle of organization which facilitates broad Lamarckian-like adaptations by improvisation. *Biology Direct* 10:68.

The detailed molecular mechanisms by which bacteria can achieve rapid evolution in response to environmental stress have been revealed by Bos, J., Q. Zhang, S. Vyawahare, *et al.* (2015) Emergence of antibiotic resistance from multinucleated bacterial filaments. *PNAS* 112:178–183. The SOS response to an antibiotic, ciprofloxacin, consists in creating multiple chromosomal filaments, which may then evolve via recombinant transfer of DNA between the multiple chromosomes: 'the strategy of generating multiple mutant chromosomes within a single cell may represent a widespread and conserved mechanism for the rapid evolution of genome change in response to unfavorable environments (i.e., chemo-therapy drugs and antibiotics)'.

Louise Johnson's lab in Reading University has shown that bacteria that have lost their flagella through deletion of the relevant DNA sequence can evolve the regulatory networks required to restore flagella and so restore motility in response to a stressful environment within just four days. Taylor, T.B., G. Mulley, A.H. Dills, *et al.* (2015) Evolutionary resurrection of flagellar motility via rewiring of the nitrogen regulation system. *Science* 347:1014–1017. The editor's summary reads: 'Two stereotypical mutations diverted an evolutionarily related regulator that normally controls nitrogen uptake to control flagella biosynthesis. The mutations increased the levels of the co-opted regulator, then altered its specificity for the flagella pathway.'

10 Bender, W. and A. Hudson (2000) P element homing to the *Drosophila* bithorax complex. *Development* 127:3981–3992.

11 It is often said that Crick used the word 'dogma' more as a joke than in all seriousness. Even if that is true, many biologists and popularisers have interpreted it very seriously.

12 For the details on Popper's lecture, see Niemann, H.J. (2014) *Karl Popper and the Two New Secrets of Life* (Mohr Siebeck, Tübingen). Noble, D. (2014) Secrets of life from beyond the grave. *Physiology News* 97:34–35.

13 McClintock, B. (1984) The significance of responses of the genome to challenge. *Science* 226:792–801.

14 Transposing long sequences breaks the dogma of genetic change being purely random since these can be sequences that *already* have functional significance with a greater probability than chance. Novelty in the evolutionary process does not therefore have to wait for gradual accumulation of random point mutations. Imagine a child playing with a construction toy. If all the child has is a pile of small bricks, then it will have to create whatever it wishes to make by small accretions. But if the child has a collection of already-formed shapes the construction can be much faster and more guided by whatever already exists. This is the process of conditioned arising. In the evolutionary process it must greatly increase the chances of functional novelty arising.

15 Mattick, J.S. (2012) Rocking the foundations of molecular genetics. *PNAS* 109:16400–16401. Recall also John Maynard's Smith's 1998 remark: 'it is not clear why he thought it [Weismann's claim that the germ line is independent of the soma] was true' (see the notes to Chapter 5).

16 Lander, E.S., L.M. Linton, B. Birren, *et al.* (2001) Initial sequencing and analysis of the human genome. *Nature* 409:860–921 – figure 42. www.nature.com/nature/journal/v409/n6822/fig_tab/409860a0_F42 .html. The domains are SET, a chromatin protein methyltransferase domain; SWI2, a superfamily II helicase/ATPase domain; Sa, sant domain; Br, bromo domain; Ch, chromodomain; C, a cysteine triad motif associated with the Msl-2 and SET domains; A, AT hook motif; EP1/EP2, enhancer of polycomb domains 1 and 2; Znf, zinc finger; sja, SET-JOR-associated domain (L. Aravind, unpublished); Me, DNA methylase/Hrx-associated DNA binding zinc finger; Ba, bromo-associated homology motif. The chromatin proteins are indicated below each example in the original *Nature* paper and can

be found in the UniProt Knowledgebase: www.uniprot.org/help/
uniprotkb.

17 'Here we show that signaling pathways that sense environmental
nutrients control genome change at the ribosomal DNA. This
demonstrates that not all genome changes occur at random and that
*cells possess specific mechanisms to optimize their genome in response
to the environment.*' My italics. See Jack, C.V., C. Cruz, R.M. Hull,
et al. (2015) Regulation of ribosomal DNA amplification by the TOR
pathway. *PNAS*: www.pnas.org/cgi/doi/10.1073/pnas.1505015112

18 There are various ways in which organisms might harness chance
variations to achieve functional inherited change. Yoav Soen has
analysed such a process, which depends on feedback between stress
and mutation rates, similar to that employed by the immune system.
Favourable mutations would reduce stress and vice versa. This would
automatically lead to functional change. Soen, Y., M. Knafo and M.
Elgart (2015) A principle of organisation which facilitates broad
Lamarckian-like adaptations by improvisation. *Biology Direct* 10:68.
This is also a good example of stochasticity at low (molecular) levels
'hiding' functional change at higher levels. Stress is a systems
property of the whole organism and would not be 'seen' at the
molecular level. Soen *et al.* also show how the hypothesis can be
tested experimentally. See also note 10 above.

19 Margulis did, however, freely acknowledge her antecedents,
including notably the Russian scientist Constantine Mereschowsky,
who argued as early as 1905 that chloroplasts derived from
cyanobacteria, while another Russian, Vladimir Kozo-Polyanski,
later extended the idea to mitochondria.

20 www.voicesfromoxford.org/news/margulisdawkins-debate/158

21 For an excellent account of these stages, see Brasier, M. (2012) *Secret
Chambers: The Inside Story of Cells and Complex Life* (Oxford
University Press, Oxford). Brasier argues that it could even have
taken a billion years prior to the Cambrian explosion for full
integration to occur, with the great majority of the organelle DNA
moving to the nucleus.

22 Although only a few genes remain in the mitochondria, the extensive
influence they have in interaction with the nuclear DNA is complex
and surprising. Hamilton, G. (2015) The mitochondria mystery.
Nature 525:444–446. See also Hutter, C.M. and D.M. Rand (1995)

Competition between mitochondrial haplotypes in distinct nuclear genetic environments: *Drosophila pseudoobscura* vs. *D. persimilis*. *Genetics* 140:537–548.

23 http://musicoflife.co.uk/pdfs/HOMAGE_COMMENTARY_Music %20of%20Life.pdf

24 A valid scientific theory must be falsifiable. A theory that is elastic enough to explain any new experimental observation is not a valid scientific theory. It is a conceptual belief system.

25 The process can involve multiple genomes. A remarkable example is the development of the eye-like ocelloid in dinoflagellates, where the components of a camera-like eye have been shown to derive from different organelles. The cornea-like structure derives from mitochondria, while the retinal structure derives from plastids, originating from symbiosis with red algae. Gavelis, G.S., S. Hayakawa, R.A. White, *et al.* (2015) Eye-like ocelloids are built from different endosymbiotically acquired components. *Nature* 523:204–207.

26 Acuna, R., B.E. Padilla, C.P. Florez-Ramos, *et al.* (2012) Adaptive horizontal transfer of a bacterial gene to an invasive insect pest of coffee. *PNAS* 109:4197–4202.

27 Hespeels, B., X. Li, J.-F. Flot, *et al.* (2015) Against all odds: trehalose-6-phosphate synthase and trehalase genes in the bdelloid rotifer *Adineta vaga* were acquired by horizontal gene transfer and are upregulated during desiccation. *PLoS ONE* 10(7):e0131313.

28 Schwartz, J.A., N.E. Curtis and S.K. Pierce (2014) FISH labelling reveals a horizontally transferred algal (*Vaucheria litorea*) nuclear gene on a sea slug (*Elysia chlorotica*) chromosome. *Biological Bulletin* 227:300–312.

29 See Enders, Giulia (2015) *Gut: The Inside Story of Our Body's Most Underrated Organ* (Greystone Books, Vancouver). An example of non-Mendelian inheritance attributable to gut bacteria is found in: Fridmann-Sirkis, Y., S. Stern, M. Elgart, *et al.* (2014) Delayed development induced by toxicity to the host can be inherited by a bacterial-dependent, transgenerational effect. *Frontiers in Genetics* 5; DOI:10.3389/fgene.2014.00027.

30 Moon, C., M.T. Baldridge, M.A. Wallace, *et al.* (2015) Vertically transmitted faecal IgA levels determine extra-chromosomal phenotypic variation. *Nature*. DOI:10.1038/nature14139.

Velasquez-Manoff, M. (2015) Gut microbiome: the peacekeepers. *Nature* 518:S3–S11. Fridmann-Sirkis *et al.* Delayed development induced by toxicity. Soen, Y. (2014) Environmental disruption of host–microbe co-adaptation as a potential driving force in evolution. *Frontiers in Genetics* 5; DOI: 10.3389/fgene.2014.00168.

31 See Nowak, M. and R. Highfield (2012) *Supercooperators: Beyond the Survival of the Fittest: Why Cooperation, Not Competition, is the Key to Life* (Simon and Schuster, New York).

8

Epigenetics and a Relativistic Theory of Evolution

The progressive triumph of physiology over molecular biology.
Sir James Black, Nobel Prize winner for drug discovery,
The Logic of Life, 1993.

Epigenetics Viewed from Physiological High Ground

I came into epigenetics by the back door. It often happens in scientific research that you can never know where an unexpected finding will lead you. My research team in 1979 was working on the natural rhythm generator of the heart: a small strip of tissue called the sinus node. It sends regular waves of excitation to the rest of the heart, which then causes it to beat rhythmically. Our earlier work had identified a rhythm-generating channel in another region of the heart. We were looking for the same channel in the sinus node. A young Italian scientist, Dario DiFrancesco, was working with my wife, Susan Noble, and a colleague, Hilary Brown, to perform the difficult dissections and recordings. The result was a paper in *Nature*[1] identifying the current through the channel, which they called i_f (following the lab jargon which dubbed it the 'funny current'), and that it was increased by the heart's natural accelerator, adrenaline. Why 'funny', and why did this bit of lab jargon become its usual name? There was a niggling doubt about its origin. It was called 'funny' because of that doubt. They wrote 'we have therefore been unable to determine whether it has a reversal potential close to the equilibrium potential for potassium ions'. The reader does not need to understand the detail of this sentence, other than to know that a potassium channel *must* have this.[2]

This little gap in the evidence widened into a huge chasm when, a year later, Dario proved it to be a different class of protein channel altogether.[3] For the next five years he and I worked to completely reformulate the mathematical theory of cardiac rhythm. The result was a seminal paper in 1985 that has formed the basis of this field ever since.[4] The mathematics of heart rhythm has now become a huge field of research called the Cardiac Physiome.[5] The paper was featured by the Royal Society in 2015 as one of a select group of papers to celebrate 350 years of publication of *Philosophical Transactions*. We used it immediately in 1985 to derive an early model of rhythm in the sinus node.[6]

There were two very important consequences of that work.

First, heart rhythm is a multiple fail-safe mechanism. Several different networks of interactions can generate the rhythm, even when the most important one is completely removed, either by a blocking drug or by genetic manipulation (a knockout). In particular, block of the i_f channel would be expected to gently slow the heart, not arrest it. This insight led to the development by a French pharmaceutical company, Servier, of a drug called ivabradine that is now used to treat patients for whom a frequency limiter of this kind is necessary. Thousands of lives must already have been saved. For the scientific basis of this development Dario DiFrancesco was awarded a prestigious prize by the French Academy of Sciences.[7]

This example of drug discovery also follows the successful strategy used by Sir James Black in his Nobel Prize work on beta-blockers and H2 receptor antagonists.[8] The method is to drill down from a high-level physiological function to identify proteins that play a relevant role, and then to the DNA that forms the template. The reverse process starting with the DNA database is like looking for needles in haystacks.[9] The correlations between DNA and function are too weak for that to be a successful strategy, as shown by the fact that 15 years after the first complete draft of the human genome the number of successful clinical applications is far from fulfilling the dream.[10]

Second, this discovery is a practical illustration of the diagram in Chapter 5 showing the primary role of the networks in relating genes to function. Networks form a screen through which it is difficult for us to work out individual contributions of the many genes that contribute their products to each function. In fact, it is sometimes impossible without insight into those systems of interactions. In a systematic study of knockouts in the unicellular organism yeast, 80% of knockouts were found to be silent.[11] It requires systems analysis of the response of the organism to various forms of environmental stress to work out what is happening. The general answer is that the networks can canalise the response towards different functional processes. We shall discuss the meaning of the word 'canalise' below, where I will also explain that canalisation of development and of reactions to the environment is precisely the original definition of epigenetics.

Epigenetic and Other Lamarckian Inheritance

Conrad Waddington's Experiments

Epigenetics means 'above genetics' and it was originally conceived by the great developmental biologist Conrad Waddington to describe the existence of mechanisms of inheritance in addition to (over and above) standard genetics. Waddington regarded himself as a Darwinist but, significantly, not a Neo-Darwinist. He was a profound thinker about biology, and much else too. His 1957 book *The Strategy of the Genes* is a masterly account of the many reasons for which he dissented from Neo-Darwinism, and it has stood the test of time. It was reprinted over half a century later, in 2014. Many of his experiments concerned the inheritance of acquired characteristics, and he discovered several processes by which it could occur. He did not describe himself as a Lamarckian, but by revealing mechanisms of inheritance of acquired characteristics I think he should be (see glossary item on Lamarckism). The reason he did not do so is that Lamarck could not have conceived of the processes that Waddington revealed. That is correct, and it is also true to say that Lamarck did not invent the idea of the inheritance of acquired characteristics. But, whether historically correct or not, we are stuck today with the term Lamarckian for this kind of inheritance.

Why did Waddington succeed experimentally where Weismann failed? Waddington realised that the way to succeed in testing for the inheritance of acquired characteristics is first to discover what forms of developmental plasticity already exist in a population, or that the population could be persuaded to demonstrate with a little nudging from the environment. This approach is more finely nuanced than using surgery since it is playing into plasticity that is already present. Surgery clearly does not achieve that since it is extremely unlikely that natural plasticity should respond to non-functional surgery. By exploiting plasticity that already existed, Waddington was much more likely to mimic a path that evolution itself could have taken.

He used the word 'canalised' for this kind of persuasion since he represented the developmental process as a series of 'decisions' that could be represented as 'valleys' and 'forks' in a developmental landscape. He knew from his developmental studies that embryo fruit flies could be persuaded to show different wing structure simply by changing the environmental temperature or by a chemical stimulus. In the developmental landscape this could be represented as a small manipulation in slope that would lead to favouring one channel in the landscape rather than another, so that the adult could show a different phenotype starting from the same genotype.

The next step in his experiment was to select for and breed from the animals that displayed the new characteristic. Exposed to the same environmental stimulus these gave rise to progeny with an even higher proportion of adults displaying the new character. After around 14 generations he found that he could then breed from the animals and obtain robust inheritance of the new character even without applying the environmental stimulus.[12] The characteristic had therefore become locked into the genetics of the animal. He called this process genetic assimilation. The full significance of this process of assimilation will become clearer in a later section on the origin of species.

Orthodox Neo-Darwinists dismissed Waddington's findings as merely an example of the evolution of phenotype plasticity. That is what you will find in many of the biology textbooks. I think that is to misrepresent what Waddington showed. Of course, plasticity can evolve, and that itself could be by a Neo-Darwinist or Darwinist or any other mechanism. But Waddington was not simply showing the evolution of plasticity *in general*; he was showing how it could be exploited to enable a *particular* acquired characteristic *in response to an environmental change* to be inherited and become assimilated into the genome. To repeat: the characteristic was acquired as a result of an environmental change, and it was inherited. That is the definition of Lamarckian inheritance that

I and many others now use. But the designation doesn't really matter. What does is that he discovered a protocol by which an acquired characteristic could be inherited. Evolution could have used the same kind of protocol.

What was happening at the gene level in Waddington's experiments? One explanation might be that some mutations occurred. That is possible but unlikely on the time scale of the experiments, which is just a few generations. Remember that the Neo-Darwinist idea is that slow accumulation of many small mutations eventually gets selected over many generations. Moreover, random mutations would occur in individuals, not in a whole group. Even if the correct combination of mutations did occur, they would have to be ones that favoured the assimilation of the characteristic. That would be a Lamarckian combination, a change that responds to the environmental change in a functional way. It is difficult to see how that could have occurred. Single small mutations would have taken many generations to spread through whole populations, and many such mutations would have been required.

But I think there is a much simpler explanation. Recall that the experiments exploited plasticity that is already present in the population. That strongly suggests that all the alleles (gene variants) necessary for the inheritance of the characteristic were already present in the population, but not initially in any particular individuals in the correct combination. The experiment simply brings them together. This is a modification of the pattern of the genome in response to the environmental change, but not in a way that requires any new mutations.

I came to this conclusion before reading *The Strategy of the Genes*. But it is in fact one of Waddington's own ideas! He writes 'There is no . . . reason which would prevent us from imagining that all the genes which eventually make up the assimilated genotype were already present in the population before the selection began, and only required bringing together.'[13] Not only does he clearly see this possibility, he also tests it. He continues: 'Attempts to carry out genetic assimilation starting from inbred lines (that is, stocks with no genetic variability) have remained quite unsuccessful. This provides further evidence that the process depends on the utilisation of genetic variability in the foundation stock with which the experiment begins.'[14] His text could not be clearer. Yet his work was sidelined. I find it shameful that this kind of denigration can be done in the name of science, just as it is shameful that Lamarck is not recognised for what he achieved. These are the reasons why the dog-matism of Neo-Darwinist popularisation has been damaging. What was missed was the opportunity to develop an integrated, and what I would call a relativistic, theory of evolution over 50 years ago. That is a long time for an important lead to be ignored.

It would have been an integrated theory because it postulates multiple mechanisms requiring a systems approach. It would have been relativis-tic because it would have been multi-scale, including processes that do not depend on a genes-eye view. Half a century later we are only now beginning to put the pieces back together again into such a synthesis. Of course, such an integrated relativistic theory builds on the undoubted successes and insights of Neo-Darwinism, just as relativity theory in physics builds on Newtonian mechanics. To repeat: Neo-Darwinism is not so much wrong as incomplete. So was mechanics when quantum mechanics appeared.

Modern Epigenetic Mechanisms

The term epigenetics still refers to influences 'above' the genome, but has acquired an additional meaning today which Waddington did not fore-see.[15] The additional meaning refers to control of the genome by marking it, or the proteins called histones around which it is wound, by means of chemical modification. These marks change and so control the expres-sion levels of genes.

The idea of the genome being controlled by the rest of the organism is not, however, new. It has been known for a long time that gene expres-sion is controlled by proteins called transcription factors that can bind to particular regions of DNA that function as switches in the sense that binding to them can switch genes on or off. The profile of these transcrip-tion factors forms a kind of epigenetics since it determines the overall pattern of gene expression, which in turn determines what kind of cell develops. More usually, epigenetics refers to additional control mecha-nisms, including marking using methylation of some of the nucleic acids, and binding to the tails of the histone proteins. When this kind of epi-genetics was first discovered it was thought that these chemical marks were always removed in the germ line before transmission to the next generation. More recent experiments have shown that this is not always true. Moreover, in some cases it has been found that the marks not only persist across generations but that the inheritance can sometimes be just as strong as standard genetic inheritance.[16]

There is now an extensive and rapidly growing literature on the inheritance of epigenetic variations.[17] In this section I will use just four examples to illustrate the very wide nature of the mechanisms involved. They represent different ways in which nature has exploited leaks in the Weismann Barrier, discussed in Chapter 7.

The tiny planarian worm, *C. elegans*, is a favourite organism for genetic and molecular biological studies. It can be infected with a particular virus. Organisms that possess the correct DNA can react to this environmental stimulus by making an RNA that silences the virus, preventing it from using the host mechanisms for reproduction. By breeding these worms with others that do not have the relevant DNA, Oded Rechavi and his colleagues obtained worms in subsequent generations that do not have the relevant DNA. Yet they still inherit the acquired resistance to the virus.[18] They do so by small quantities of the viral-silencing RNA passing through the male germ line to be amplified in each generation by an enzyme called RNA polymerase. The acquired characteristic is transmitted in this way through at least 100 generations. This example shows that the idea that an acquired characteristic will necessarily die out after a few generations is not correct. It also reveals that RNAs can also be transmitted through the germ line. DNA is not the only inherited material.

Robust inheritance of an acquired epigenetic characteristic has been demonstrated in mice by Joe Nadeau's group in Seattle. They worked on a family of proteins that can insert mutations in DNA and RNA to show inheritance of epigenetic marking. This shows that the genome is not completely wiped clean of the marks in the germ line. On the contrary, Nadeau's work shows that such inheritance can be just as robust as standard genetic inheritance and can persist for many generations.[19] Epigenetic marking of the chromosome proteins has also been shown to be inherited.[20] The transmission of epigenetic marking has recently been shown to play a role in the inheritance of obesity in humans,[21] while the transmission of RNAs in sperm mediates the transmission of obesity in mice.[22]

Epigenetic mechanisms seem to be able to transmit memories of unpleasant experiences. Kerry Ressler and Brian Dias at Emory University in the USA have shown that mice can be trained to fear a particular chemical smell through association of the smell with an electric shock. The progeny display the same fear of the smell even though they were not trained to do so. The precise epigenetic mechanism in this case remains to be discovered.[23]

Another mechanism by which evolution can use epigenetics to bypass the Weismann Barrier is to transmit the epigenetic marks through behaviour. This process has been demonstrated by Michael Meaney's group in Canada. Rodents, like many other animals, groom their young by licking and stroking them. This behaviour enhances the health and longevity of the progeny. It also influences epigenetic marking in the region of the brain called the hippocampus which, amongst other roles, plays a part in emotional behaviour. The epigenetic effects can therefore predispose the progeny to show the same behaviour towards their young. This form of epigenetic inheritance doesn't even require transmission through the germ line. It is a behavioural way of bypassing the Weismann Barrier.[24]

These examples suffice to show what is happening in modern epigenetic research.[25] Can these and many other examples be dismissed as the rare exceptions that, to use a common phrase, 'prove the rule'? Could the Neo-Darwinist synthesis live with that? After all, Newtonian physics lives on despite the exceptions at the micro scale of quantum mechanics and the mega scale of general relativity. Those exceptions really are negligible for the spatial and time scales at which the physics of everyday life operates. We continue to use Newton's equations successfully. In the case of the exceptions to standard inheritance in evolutionary theory, this option is not open to us precisely because those exceptions operate at the same spatial scales and over the same time periods as the evolutionary process itself. Inheritance is inheritance, whether it is genetic, epigenetic, RNA-based, culturally based or whatever. All the inheritance processes end up doing the same thing, which is to modify the organism. They are intricately interconnected so that selection would not be able to distinguish between them. They also interact so much that disentangling the genetic and epigenetic processes may be the wrong way to look at the situation.[26]

But aren't the non-standard mechanisms rare? That is a good question and the best answer at this early stage in research on epigenetic inheritance is that we don't know how rare it may be compared to mutations in DNA. But rarity is not really the issue. Speciation is also rare. Thousands of years of selection of dogs, cats and fish have not resulted in new species by the standard definitions, such as whether or not the variants can interbreed. Moreover, as we will see in a later section of this chapter when we look at research on Darwin's finches, epigenetic variation can be shown to be just as frequent as or even more frequent than genetic variation.

Niche Construction and the Active Role of Organisms

We saw in the previous chapter that Karl Popper's 1986 lecture to the Royal Society anticipated the significance for evolutionary theory of the discovery of reverse transcription. Popper's lecture also contained a second major criticism of Neo-Darwinism.

This was based on his clear understanding of a phenomenon usually known as the 'Baldwin effect'[27] or 'adaptability driver'.[28] Organisms can choose new niches for themselves and their descendants.[29] Moving to a new niche can change the course of evolution even with no mutations whatsoever. That choice is a physiological characteristic of the phenotype, not a change in DNA. So how can it change the course of evolution? The answer is surprisingly simple. In a wild population in which individual genomes are not identical, the combinations of alleles in the adventurous organisms discovering new niches will be favoured. That is an evolution of the genome by combinatorial selection, not selection of new random mutations.

It is not surprising that a logician like Popper should have immediately understood the immense significance of this fact. To illustrate it he even invented an imaginary world in which there was no competition for survival, no 'selfish genes'. The organisms would still evolve. Of course, the world in which such evolution could occur would have to be effectively infinite in size to accommodate all the organisms that would develop. But this was just a thought experiment familiar to scientific theorists and philosophers. Conrad Waddington understood the same point.

Why, then, do selfish gene theorists ignore it? They do so by taking an atomistic gene-centred view. From that viewpoint no new genes developed and the process can be represented as not very different from random drift within an existing gene pool, which I will discuss later in this chapter. From a physiological or psychological functional viewpoint, however, the process is an active choice of the organisms, including learnt behaviour. The causality involved here is very far from random.[30] Once the new niche choice has been made, the functionality can become assimilated into the genome just as in Waddington's experiments. As Popper also saw, it is the insistence on the gene-centred approach that is the problem. By contrast, it is combinations of genes, or rather combinatorial interactions between large numbers of their products, RNAs and proteins, that are important functionally. What may appear to be a random distribution of components at a molecular level may in fact represent the particular pattern that has functional significance at a higher level.

This is a very important and little-understood consequence of taking a biological relativistic view. Interactions at a high level, which can be clearly seen as functional, may not appear to be functional at a low level. This is one of the reasons why Neo-Darwinists can use a gene-centric view to claim that all variation is random with respect to function. The non-randomness may only be evident if one takes a high-level perspective.

In addition to the fact that the functional pattern may not be visible at a molecular level, most single genes contribute very little to complex functions, which is why the correlations between genes and complex diseases have been found to be a matter of large numbers of very small effects, still summing up to a small overall fraction of causation. The atomistic view was never going to be of much use in physiology and pathology. This is the point made by Sir James Black in the quotation at the head of this chapter.

There is another reason why Popper was ignored. He never published his Royal Society lecture! When he died at the age of 92 he was still in correspondence with the Nobel laureate and director of the Cambridge Molecular Biology laboratory Max Perutz, who he hoped to convince before he published. The lecture was then locked up for 35 years in the Popper archive in the USA. Fortunately, it has now been released and has been published as part of a book, *Karl Popper and the Two New Secrets of Life*.

Niche construction and the adaptability driver are examples of natural purposiveness. Organisms actively seek and create new niches.[31]

While standard evolutionary biology ignores the deep significance of the adaptability driver,[32] it has long acknowledged a related but much weaker form of the idea. This is the idea of genetic drift. Even without a new niche the offspring of organisms will not necessarily display the same distribution of gene variants in their population as their parents. Over time, even a random process can lead to change.[33] The adaptability driver adds the important consequence of some members of a population *actively* choosing or creating a new niche. The causation is at the level of the phenotype, not the genotype.

The Origin of Species?

Despite the title of Darwin's famous book, we cannot be sure to what extent natural selection acting on gradual accumulation of small mutations has accounted for the development of new species. Many of the examples that are often quoted as demonstrating that this is the main source of new species show that new species have arisen in new niches, e.g. by geographic isolation as in the case of the Galapagos Island tortoises and finches. But this evidence, by itself, does not establish what the mechanism may have been. Other interpretations would also be possible.

Darwin's idea of natural selection was in part based on observing the effects of artificial selection, producing many different varieties of dogs, cats, fish, etc. But it is important to note that thousands of years of this kind of selection has produced new varieties, not fully fledged new species as defined by inability to interbreed. At best, this demonstrates selection's ability to produce incipient development of new species.

One of the best examples of species development is the variety of birds called greenish warblers, forming what is called a ring species around the Himalayas (Figure 8.1).

North of the Himalayas there are two varieties (subspecies?) shown as blue and red in Figure 8.1 that co-exist but do not interbreed. But they are each connected to varieties with which they do interbreed, stretching around the Himalayas to join the ring together in the south. Claud Ticehurst proposed in 1938 that the greenish warblers started in the south, then slowly evolved as they spread around the Himalayas to the west and east, to eventually meet in the north when they had become different species.[34] Very recent genome studies have shown that 'although spatial patterns of genetic variation are currently mostly as expected of a

Figure 8.1 Presumed evolution of greenish warblers around the Himalayas. The colours indicate the subspecies: yellow: *P. t. trochiloides*; orange: *P. t. obscuratus*; red: *P. t. plumbeitarsus*; green: *P. t. 'ludlowi'*; blue: *P. t. viridanus* (https://commons.wikimedia. org/wiki/File:Greenish_warbler_ring.svg; Public Domain).

ring species, historical breaks in gene flow have existed at more than one location around the ring, and the two Siberian forms have occasionally interbred'.[35]

This is perhaps the best example we have of this process, which itself is rare.[36] The new evidence supports and extends Ticehurst's hypothesis about the historical development, since the two northern forms are the most distinct genetically, but it does not in itself prove the precise mechanism. It does not, for example, exclude an influence of the environment on the selection of combinations of genes, or any other process by which a Lamarckian process becomes assimilated into the genome via environmental influence on epigenetic variation. Nor does it exclude natural genetic engineering. This is a difficulty with any study of existing genetic variations to infer the process by which those variations originally arose.

It is important to emphasise that the assumption that genetic variations responsible for the origin of species are independent of the organism and its environment is just that: an assumption. It has not been proven experimentally to produce new species. By contrast, hybridisation in plants readily generates new species, and the process of symbiogenesis must also have done so, particularly when eukaryotes evolved from prokaryotes. Natural genetic engineering could also do so if it involves DNA shuffling that removes reproductive compatibility.

The important message here is not that the generation of new species by natural selection working on random mutations is impossible, but rather that the examples often quoted to prove that it happened are not so clear-cut when they are examined closely. We have to distinguish evidence about the *historical* development of new species from evidence about the precise *mechanisms* by which this happened. Moreover, as emphasised earlier in this chapter, the different mechanisms might be so intertwined that it may be inappropriate to think in terms of single mechanisms.

Can we obtain evidence from genetic and epigenetic data that such processes involving both epigenetic variation and genetic assimilation may have occurred? Experiments to answer this question were performed very recently. Without doubt, the original icons of Darwinian evolution are the tortoises and finches of the Galapagos Islands, the organisms that inspired Darwin's work following his voyage on *The Beagle*. Darwin would not have been able to know what caused the variations, but we can now begin to do so.

Michael Skinner's research team in the USA has studied the epigenetic and genetic variations in these finches.[37] Figure 8.2 summarises the results of their work in terms of the number of genetic (blue) and epigenetic (red) variations (mutations) in four of the various finch species compared to the reference species, *Geospiza fortis*. The results are consistent with the view that both epigenetic and genetic variation have occurred and that the number of both kinds of mutations increases with the presumed evolutionary distance between the species. The smallest numbers, 84 epimutations and 34 genetic mutations, are found between the closest species, while the largest, 1062 epimutations and 602 genetic mutations, are found between the most distant species. Moreover, epigenetic changes in DNA methylation correlated better with evolutionary distance than DNA sequence changes, and many of the epigenetic changes clustered around genes involved in bone morphogenesis.

This is what we might expect if the initial variation in beak formation had been epigenetic followed by genetic assimilation, or more likely the

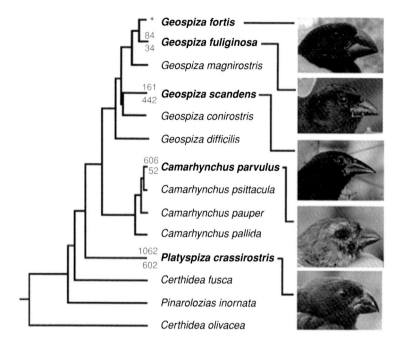

* – Reference species
Red – Epimutations (DMR)
Blue – Genetic mutations (CNV)

Figure 8.2 Genetic and epigenetic variation in the finches of the Galapagos Islands. The tree diagram on the left shows the relatedness of the 14 individual species. Four of these species were compared with the one chosen as the reference species, *Geospiza fortis*. DMR = DNA methylation region; CNV = copy number variation. The number of epigenetic mutations is shown as red, the number of genetic mutations in blue (from Skinner *et al.* 2014).[38] For a colour version of this figure, please see the plate section.

epigenetic and genetic changes simply intertwined rather than occurring sequentially. As Skinner and his team conclude, 'epigenetic and genetic changes may jointly regulate genome activity and evolution, as recent evolutionary biology modeling suggests'. What is clear from these studies is that automatically assuming genetic mutation as the sole driver of evolutionary change is no longer secure as an explanation. Genes as DNA sequences may also be followers rather than drivers. That would match well with what happens in all cells: the cell networks tell the genome what to do, not the other way round.[39]

Are Genes Followers Rather than Leaders?

How does Neo-Darwinism respond to this kind of explanation? The answers usually illustrate the confusion that we found in Chapter 5 over the definition and nature of genes. Consider the response to the question raised in an article on genes as followers in the *New Scientist* (12 October 2013) which quotes Richard Dawkins: 'which elements have the property that variations in them are replicated with the type of fidelity that potentially carries them through an indefinitely large number of evolutionary generations?'. Dawkins answers his own question: 'Genes certainly meet the criterion. If anything else does, let's hear it.'[40] But he doesn't clarify which definition of a gene he is using. If such an answer refers to DNA sequences, as I suspect it does in this case, then it is clearly true that much more is inherited. Moreover, as we have seen, DNA is not the only replicator. Cells also replicate, and it is only through their replication via self-templating that DNA can replicate. The two, DNA and the rest of the cell, always and must replicate together. The famed 'immortality' of DNA is actually a property of cells. Only cells have the machinery to correct the frequent faults that occur in DNA replication.

There is a common misunderstanding of what counts as experimental evidence for Neo-Darwinism. Evidence for evolutionary change and for speciation is now so extensive as to be virtually incontrovertible. It is often taken to be evidence for Neo-Darwinism as a specific theory of how it all happened. But that evidence is perfectly consistent with other mechanisms outside the standard framework of Neo-Darwinism.

Given the extent to which genomic reorganisation has occurred during evolution, which was the subject of Chapter 7, and the extent to which the genome is controlled by epigenetic mechanisms, which can in turn be assimilated into the genome, it makes much more sense to see the biological network processes as the driver. The genes, defined as sequences of DNA, dance to the tune of evolution, to the tune of life itself.

Isn't a Lot of DNA 'Selfish', 'Parasitic'?

One of the standard defences of selfish gene theory is based on the discovery that, in humans and many other organisms, only a few per cent of the genome codes for proteins and can therefore be classified as 'genes' in the original molecular biological sense. The rest was described as 'junk' DNA, the ultimate example of 'selfishness' since it was seen as DNA 'hitching a ride' with no function, a bit like a virus that has become permanently resident in the body. The strong implication is that this discovery favours the selfish gene view.

We now know that is a confusing way of viewing genomes. There are several ways in which the confusion can be unravelled.

The words 'selfish', 'junk', etc. are, of course, metaphors. More importantly they are empirically empty metaphors when applied to sequences of DNA. No conceivable experiment could validate or invalidate them, as explained in Chapter 5. That the metaphors are empty from the standpoint of empirical science does not mean they have no impact. On the contrary, they have had, and still have, very persuasive impact on the way in which many people think about biology.

A more persuasive counter-argument is therefore needed. Recent experimental work has provided us with precisely that. The more we examine non-protein-coding DNA the more evidence we find that an overwhelming 80% is transcribed to form RNAs, and that around 20% of those are already known to have function.[41] It was far too premature to dismiss the great majority of the human genome as junk.[42]

This also raises questions about what precisely is a gene. In the original sense defined by Johanssen in 1909 (see Chapter 5) a gene is anything that contributes to inheritance. At the least, therefore, all the DNA that does that should be regarded as genes. It is too restrictive to define 'gene' as only protein-coding DNA. On a relativistic view we should, however, go even further than that. As we have seen from Waddington's and other experiments, the whole pattern of the genome is relevant to inheritance. The real problem with the gene-centred view is that it ignores the properties of the genome as a whole. These overall patterns may have functional significance through the eyes of a higher-level analysis at which functional organisation and processes appear.

The Speed of Evolution

Some of the problems in evolutionary theory concern the speed of speciation. It is clear from the geological record that this has not always been smooth. One of the reasons why Cuvier disagreed with Lamarck in the early nineteenth century is that he was able to discredit Lamarck's idea of the transformation of species by pointing to the fact that the fossil evidence, as it was known then, was also consistent with multiple periods of creation. The record was that patchy, and it was also clear that some species remained essentially unchanged for very long periods of time, a phenomenon we call stasis. In the twentieth century, with much more evidence to consider, Niles Eldredge and Stephen Jay Gould in 1971 proposed the theory of punctuated equilibrium to account for the fact that most fossil species show long periods of stasis, and that rapid (on a geological time scale) changes occurred more rarely and were important periods of speciation.[43] The difference between their theory and that of Cuvier was that Eldredge and Gould were simply proposing that evolutionary change from a common ancestor does not happen at a constant speed, whereas Cuvier interpreted the evidence to show that there had been multiple creations.

Darwin also realised that evolutionary change could not always have been smooth. In the fourth edition of *The Origin of Species* he wrote 'the periods during which species have undergone modification, though long as measured in years, have probably been short in comparison with the periods during which they retain the same form'. This is, in essence, Eldredge and Gould's idea of punctuated equilibrium.

There has been much argument about what precisely is meant by 'punctuated'. A gradual change over 100,000, one million or even a few million years will appear rapid on a geological time scale of hundreds of millions of years. The Cambrian explosion that occurred over 500 million years ago is a good example. Within just 20 million years all the phyla in existence today had developed. The standard response to theories of punctuated evolution has therefore been that they are entirely consistent with Neo-Darwinism. Changes in selection pressure due to environmental changes or geographic distribution, and the occasional catastrophic environmental change, might account for the observed variations in speed of change without supposing additional mechanisms of change.

However, the new evidence from work on symbiogenesis, the various forms of inheritance of acquired characteristics, genetic assimilation, natural genetic engineering, including genome change and reorganisation over and above the accumulation of chance mutations, changes the situation in more fundamental ways. These mechanisms resemble punctuated equilibrium theories in proposing that evolution can occur in jumps. Since these can be very sudden indeed it is best to use a word different from 'punctuated' to avoid confusion with Eldredge and Gould's theory. 'Saltatory' means jumping. The new mechanisms produce saltations of various kinds.

Symbiogenesis is the fusion of two species. The best established example of this is the bacterial origin of mitochondria and plastids and, perhaps, other organelles.[44] Clearly, this process is the ultimate in 'saltation'. It depends on processes of cellular ingestion that are natural in the feeding activity of unicellular organisms and, on an evolutionary time scale, it is therefore very rapid indeed. Of course, subsequent changes can then also occur more slowly. We know, for example, that some of the ingested DNA in what became organelles eventually moved to the nucleus in eukaryotes.

Natural genetic engineering could also occur within a single generation. Reorganisations of genomes involving duplications, deletions and insertions of long sequences would be essentially instantaneous on a geological time scale. Defenders of the Modern Synthesis have argued that speciation due to such changes, and symbiogenesis, should not be classified as punctuated equilibrium. That is correct in the sense that it was not what Eldridge and Gould had in mind. But as far as time scale is concerned such changes would be saltatory in the ordinary sense of the word. They would be even more sudden than the punctuations proposed by Eldredge and Gould.

Genetic assimilation can also occur rapidly. Waddington's mid-twentieth-century experiments showed that an induced acquired characteristic in fruit flies could become permanent (assimilated) within 14 generations. This must have represented the time required for selection for an induced characteristic to bring together in a single genome all the relevant alleles for that characteristic to be passed on to subsequent generations without the inducing environmental stimulus.

Inheritance of acquired characteristics through the persistence of epigenetic effects through successive generations can also speed up the evolutionary process. These trans-generational environmental influences should spread through a population much more rapidly since it is possible for a large fraction of the population to be subject to the same changes at the same time. There is no need to wait for a single DNA change to spread slowly through a population. The orthodox response to this mechanism is to dismiss it as transient, which it certainly is in some cases. But, as we have now seen, there are examples of such transmission over many generations and which show the same degree of strength as standard genetic transmission. Since such effects would not need to occur very frequently, the difficulty in identifying them experimentally would be perfectly understandable. Speciation itself is a rare event.

While there can be considerable uncertainty about the relative contributions of the different mechanisms to evolutionary change, two conclusions seem clear. The first is that, with a variety of mechanisms open to the evolutionary process, it should occur faster. The second is that it is probable that the relative contributions varied at different stages in evolution, and must often have interacted with each other. A systems perspective on this opens up many possibilities for a general explanation of sudden radiations, such as occurred during the Cambrian explosion.[45]

Respecting the Principle of Biological Relativity

This chapter and the previous one have revealed the extraordinary range of counterexamples to the processes assumed in Neo-Darwinism to be sufficient to explain biological evolution. The molecular 'single gene' level is not the most appropriate scale at which to look for all those processes. Many of them depend on the pattern of gene variants and how they come together in individual organisms. Evolution by changing that pattern doesn't even require new mutations. Nor do epigenetic mechanisms. The single process of gradual accumulation of small mutations followed by natural selection is clearly inadequate as the basis of a comprehensive theory.

Respecting the principle of Biological Relativity requires that we take account of mechanisms discovered at all scales and their interactions. That is the subject of the last section of this chapter.

A Biological Relativistic View of Evolution

It has been recognised for many years that the original formulation of Neo-Darwinism needed extending. As we saw in Chapter 5, it was already extended as early as 1931 by the addition of random genetic drift[46] and later in 1968 with Motoo Kimura's Neutral Theory of Molecular Evolution,[47] which can be seen as a development of random genetic drift. These extensions recognised the need to take account of processes that could produce variation and evolutionary change based on random selection within the existing gene pool in a population, which can occur even without new mutations. In this respect, they partly resemble both the adaptability driver and Waddington's genetic assimilation process described earlier in this chapter. Both of those also depend on change of combinations within the existing gene pool without requiring new mutations.

The difference is that both the adaptability driver and Waddington's genetic assimilation describe processes that are not really consistent with the spirit of Neo-Darwinism. The adaptability driver gives to organisms the creative initiative to choose and even help to create new niches. The driver, then, is clearly the active phenotype, not the passive genotype. Genetic assimilation is even more incompatible since it is a form of inheritance of acquired characteristics. That is the reason why Waddington did not accept the Neo-Darwinist view of evolution. The inheritance of acquired characteristics is more than an extension. It is a complete break.

Further extensions were introduced more recently with the realisation that strict separation between evolution and development was incorrect. This is the central feature of an important book published in 2010 entitled *Evolution: The Extended Synthesis*.[48] That book contains many details on the experimental results that lead to an extended evolutionary synthesis (EES). Figure 8.3 shows that these include around eight extensions.

The inner circle of Darwinism includes variation, inheritance and natural selection. Neo-Darwinism adds the processes of gene mutation, Mendelian inheritance, population genetics, contingency and speciation.

The extended evolutionary synthesis is then viewed as adding evolutionary developmental biology (evo-devo theory), genetic assimilation (plasticity and accommodation), the adaptability driver and related effects (such as niche construction), which we have already described in this chapter. To these are added five more. Three of these, epigenetic inheritance, genomic evolution (which includes natural genetic engineering described in the previous chapter) and multi-level selection (which naturally respects the principle of Biological Relativity), we have already discussed. The remaining ones are replicator theory and evolvability.

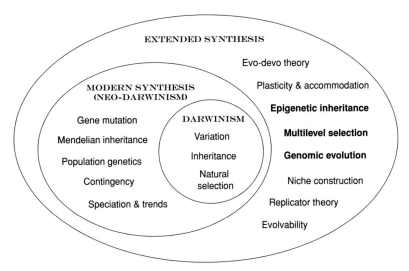

Figure 8.3 Extended evolutionary synthesis. The inner circle represents the main features of Darwinism. The inner oval represents the processes added by Neo-Darwinism. The outer oval represents the processes added in the extended evolutionary synthesis proposed by Pigliucci and Müller in 2010. I have represented three of these in bold face in anticipation of the view shown in Figure 8.4. This diagram is based on Pigliucci and Müller's diagram. Adapted from: Pigliucci, Massimo, and Gerd B. Müller, eds., *Evolution, the Extended Synthesis*, Figure 1.1, p. 11, © 2010 Massachusetts Institute of Technology, by permission of The MIT Press.

Evolvability refers to processes similar to those discussed in Chapter 7, which enable organisms to speed-up genetic variation in response to environmental influences. As we saw in that chapter, those effects can be very large and can be targeted functionally at the relevant parts of the genome.

In addition to aspects of replication already discussed in this book, replicator theory also refers to ways in which replication operates in other fields, such as chemistry and neuroscience, that are beyond our scope.[49]

I think that the evidence and ideas discussed in this book oblige us to represent the EES as a replacement because some of the features of Neo-Darwinism that I described in Chapter 5 are incompatible with what we now know. Science advances through the admission of error. In fact it makes the biggest advances when it does that. We saw in Chapter 1 how such admission of error was the process by which the conceptual foundations of physics changed at the beginning of the twentieth century.

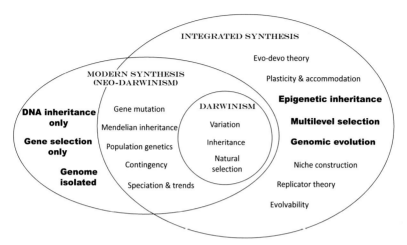

Figure 8.4 Diagram illustrating definitions of Darwinism, Modern Synthesis (Neo-Darwinism) and Integrated Synthesis. The diagram is derived from Pigliucci and Müller's (2010) presentation of an Extended Synthesis, shown in Figure 8.3. All the elements are also present in their diagram. The differences are: (1) the elements that are incompatible with the Modern Synthesis are shown in bold on the right (as they also are in Figure 8.3); (2) the reasons for the incompatibility are shown in the three corresponding bold elements on the left. These three assumptions of the Modern Synthesis lie beyond the range of what needs to extend or replace the Modern Synthesis; (3) in consequence, the Modern Synthesis is shown as an oval extending outside the range of the Extended Synthesis, which therefore becomes a replacement (from Noble 2015).[50]

I believe we have reached such a tipping point in biological science at the beginning of the twenty-first century.

Figure 8.4 illustrates the definitions and relationships between the various features of Darwinism, the Modern Synthesis and a proposed new integrative and relativistic synthesis. The diagram is based on an extension of the diagram used by Pigliucci and Müller (Figure 8.3) in explaining the idea of an EES. It can be considered as an extension of the extension to bring out the extent to which it is actually a replacement.[51] The main difference between Figure 8.3 and Figure 8.4 lies in the listing, at the left-hand side of Figure 8.4, of those tenets of the Modern Synthesis that have now become impossible to sustain.

The shift to a new synthesis in evolutionary biology can also be seen to be part of a more general shift of viewpoint within biology towards systems approaches. The reductionist approach (which inspired the Modern

Synthesis as a gene-centred theory of evolution) has been very productive, but it needs, and has always needed, to be complemented by an integrative approach that recognises causality at multiple levels. The reductionist interpretations of evolution are not wrong in the sense that the processes they include do not occur, they are simply incomplete.

In retrospect, Neo-Darwinism can therefore be seen to have oversimplified biology and over-reached itself in its rhetoric. By so conclusively excluding anything that might be interpreted as Lamarckism, it assumed what couldn't be proved. As John Maynard Smith admitted in 1998, 'it [Lamarckism] is not so obviously false as is sometimes made out',[52] a statement that is all the more significant from being made by someone working entirely within the Neo-Darwinist framework. His qualification of this statement in 1998 was that he couldn't see what the mechanism(s) might be. We can now do so thanks to the ingenious experimental research in recent years, which I have described in this and the previous chapter.

Nevertheless, the dogmatism was unnecessary and uncalled for. It damaged the reputation of Lamarck, possibly irretrievably. He was caricatured as a figure of fun. How could he be so silly as to suppose the impossible?! In fact, Lamarck should be recognised by biologists generally as one of the very first to coin and use the term 'biology' to distinguish our science, and by evolutionary biologists in particular for championing the transformation of species against some very powerful critics. Darwin praised Lamarck for this achievement: 'this justly celebrated naturalist... who upholds the doctrine that all species, including man, are descended from other species'.

Many others were damaged too, Waddington included. Waddington was a systems biologist in the full sense of the word. If we had followed his lead, many of the more naive twentieth-century popularisations of genetics and evolutionary biology could have been avoided. A little more humility in recognising the pitfalls that beset the unwary when they think they can ignore some basic philosophical principles would have been a wiser strategy. The great physicist Poincaré pointed out, in connection with the relativity principle in physics, that the worst philosophical errors are made by those who claim they are not philosophers. They do so because they don't even recognise the existence of the conceptual holes they fall into. Biology has its own version of those conceptual holes.

Conclusions

The principle of Biological Relativity has applications everywhere in biology, but there can't be much doubt that the most surprising and fundamental applications are in the field of evolutionary biology. For primarily conceptual reasons, the subject has been dominated for a whole century by a theory that made unjustifiable assumptions about what is possible in nature. Recent experimental evidence in a wide range of fields, discussed in this chapter and the previous one, have shown the need for change. But it is also true that some significant features of the new evidence were foreshadowed in work done half a century ago. Conrad Waddington led the way and should not have been ignored. So did Barbara McClintock and Lynn Margulis.

The general reader might well ask why this has taken so long when the conceptual and experimental arguments in favour of change are so convincing. Why are scientists like me apparently in such a small minority? There is a simple answer to that question. We are only *apparently* a minority.[53] I have discussed extensively with evolutionary and other biologists in the course of lecturing to audiences, large and small, all around the world. Exceedingly few of the tens of thousands involved have seriously defended the orthodox Neo-Darwinist view as a complete explanation. The more usual reaction of many biologists is 'we have known most of this for a long time', usually combined with fairly vague notions of how restrictive the original assumptions of Neo-Darwinism really are, and therefore with little understanding that Neo-Darwinism is no longer valid as the sole explanation. The vagueness is also widespread in the humanities and other sciences. The reason is that the concepts have been presented in a sufficiently elastic way that almost any new discovery can be made to seem compatible.[54] Most significantly, many do not realise that Darwin was not a Neo-Darwinist. They draw the uncomfortable conclusion that to oppose Neo-Darwinism is to oppose the theory of evolution itself. In fact, there are and always have been a number of theories on the mechanisms by which evolution happened. None of them has been conclusively shown to be the only mechanism.

So, I will finish this chapter with two important concluding reminders.

First, the evidence for evolutionary change and for the transformation of species is overwhelming. Nothing in the new experimental results and arguments for replacing Neo-Darwinist theory seriously challenges that evidence for evolutionary change. On the contrary, this evidence reinforces the conclusion that evolution has been the origin of species. The problem, as I have explained in this and previous chapters, is that the evidence for evolutionary change is consistent with, and requires, a more integrative theory that acknowledges more mechanisms of variation than the random ones assumed by Neo-Darwinism as the exclusive mechanisms.

Second, it was a central feature of Neo-Darwinism to conclusively exclude the inheritance of acquired characteristics. That even became a dogma. Dogmas have no role in science.

The view of evolution that I have developed in this book avoids that kind of dogmatic certainty. I have outlined various *possible* mechanisms, some of which are clearly outside the usual boundaries, and certainly outside the classical boundaries, of Neo-Darwinism. I have also admitted that we simply do not know which of these mechanisms contributed to or dominated evolutionary change at the various stages. The starting point in science is to admit what we don't know since that enables us to formulate the right questions. We can then design experiments and empirical observations to distinguish between the possible hypotheses.

The dance of the DNA sequences as they were first developed and then controlled by living organisms occurred in a continually unfolding theatre that has been performing for 3.5 billion years. That theatre has developed a magnificent organ – the genome – but let's not confuse the organ with the players and composers. They are life itself. The DNA nucleotides dance to life's music like puppets on strings.

Notes

1 Brown, H., D. DiFrancesco and S. Noble (1979) How does adrenaline accelerate the heart? *Nature* 280:235–236.
2 It is a necessary feature of the thermodynamics of ion channels. The relevant equation is the Nernst equation: http://en.wikipedia.org/wiki/Nernst_equation.
3 DiFrancesco, D. (1981) A new interpretation of the pace-maker current in calf Purkinje fibres. *Journal of Physiology* 314:359–376.

4 DiFrancesco, D. and D. Noble (1985) A model of cardiac electrical activity incorporating ionic pumps and concentration changes. *Philosophical Transactions of the Royal Society B* 307:353–398.

5 Noble, D., A. Garny and P. Noble (2012) How the Hodgkin–Huxley equations inspired the cardiac Physiome Project. *Journal of Physiology* 590:2613–2628. See http://physiomeproject.org and www.vph-institute.org for details on how the Physiome Project has grown.

6 Noble, D. and S.J. Noble (1984) A model of S.A. node electrical activity using a modification of the DiFrancesco–Noble (1984) equations. *Proceedings of the Royal Society B* 222:295–304.

7 The Grand Prix Lefoulon-Delalande of the Academie des Sciences was awarded to Dario DiFrancesco in 2008.

8 Black, J. (2010) A life in new drug research. *British Journal of Pharmacology* 160(suppl 1):S15–S25.

9 The immense size of the 'haystack' was discussed in Chapter 3. Interactions between gene products in the human genome could produce more possible effects than the total number of particles in the known universe.

10 An editorial in *Nature* (June 2010) spelt this out by quoting the leaders of the Human Genome Project: 'But for all the intellectual ferment of the past decade, has human health truly benefited from the sequencing of the human genome? A startlingly honest response can be found on pages 674 and 676, where the leaders of the public and private efforts, Francis Collins and Craig Venter, both say "not much".'

11 Hillenmeyer M.E., E. Fung, J. Wildenhain, *et al.* (2008) The chemical genomic portrait of yeast: uncovering a phenotype for all genes. *Science* 320:362–365. Eighty per cent of knockouts in yeast were silent in normal physiological conditions. The majority of those reveal their function in conditions of metabolic stress when the organism needs to use different networks to survive and function. Virtually all the genes are therefore functional. The normal networks must be capable of buffering most genetic changes, which is why this widely used method does not necessarily reveal function. For more details, see Noble, D. (2010) Differential and integral views of genetics in computational systems biology. *Interface Focus*, 1; DOI: 10.1098/rsfs.2010.0444.

12 Waddington, C.H. (1956) The genetic assimilation of the bithorax phenotype. *Evolution* 10:1–13. Noble, D. (2015) Conrad Waddington and the origin of epigenetics. *Journal of Experimental Biology* 218:816–818.

13 Waddington, C.H. (1957) *The Strategy of the Genes* (Allen and Unwin, London; reprinted 2014. p. 176).

14 Waddington, *The Strategy of the Genes*, p. 178.

15 Waddington's original meaning has often been forgotten in modern work on epigenetics. This is a mistake since the two kinds of epigenetic effects can interact in evolutionary processes. I agree with Patrick Bateson in continuing to use both meanings; see Bateson, P. (2014) The rise and rise of epigenetics. In *The Systems View of Life*, F. Capra, and P. Luisi, editors (Cambridge University Press, Cambridge, pp. 198–202).

16 As an example, Nelson, V.R., J.D. Heaney, P.J. Tesar, N.O. Davidson and J.H. Nadeau (2012) Transgenerational epigenetic effects of Apobec1 deficiency on testicular germ cell tumor susceptibility and embryonic viability. *PNAS* 109:E2766–E2773.

17 See the special issues on epigenetics of the *Journal of Experimental Biology* (2015), volume 218, issue 1; and *Progress in Biophysics and Molecular Biology* (2015), volume 118. Within the latter volume, see particularly the articles by Yohn *et al.*, Ozgyin *et al.*, Martos *et al.*, Ma *et al.* and Soubry.

18 Rechavi, O., G. Minevish and O. Hobert (2011) Transgenerational inheritance of an acquired small RNA-based antiviral response in *C. elegans. Cell* 147:1248–1256.

19 Nelson *et al.* Transgenerational epigenetic effects of Apobec1 Cytidine deaminase deficiency on testicular germ cell tumor susceptibility and embryonic viability. DOI: 10.1073/pnas.1207 169109.

20 McCarrey, J.R. (2015) The epigenome: a family affair. *Science* 350:634–635. Siklenka, K., S. Erkek, M. Godmann, *et al.* (2015) Disruption of histone methylation in developing sperm impairs offspring health transgenerationally. *Science* 350:aab2006.

21 Donkin, I., S. Versteyhe, L.R. Ingerslev, *et al.* (2015) Obesity and bariatric surgery drive epigenetic variation of spermatozoa in humans. *Cell Metabolism*; DOI: 10.1016/j.cmet.2015.11.004.

22 Chen, Q., M. Yan, Z. Cao, *et al.* (2015) Sperm tsRNAs contribute to intergenerational inheritance of an acquired metabolic disorder.

Science; DOI: 10.1126/science.aad7977. Sharma, U., C.C. Conine, J.M. Shea, *et al.* (2015). Biogenesis and function of tRNA fragments during sperm maturation and fertilization in mammals. *Science*; DOI: 10.1126/science.aad6780.

23 Dias, B.G. and K.J. Ressler (2014) Parental olfactory experience influences behavior and neural structure in subsequent generations. *Nature Neuroscience* 17:89–96. Dias, B.G. and K.J. Ressler (2014) Experimental evidence needed to demonstrate inter- and trans-generational effects of ancestral experiences in mammals. *Bioessays* 36:919–923. This research is ground-breaking, but needs to be confirmed in other laboratories, and the molecular mechanisms still have to be worked out.

24 Weaver, I.C.G. (2009) Life at the interface between a dynamic environment and a fixed genome. In *Mammalian Brain Development*, D. Janigro, editor (Humana Press, Springer, New York; pp. 17–40).

25 This field has exploded. See Tollefsbol, T. (ed.) (2014) *Transgenerational Epigenetics: Evidence and Debate* (Academic Press, New York). The conclusion of the opening chapter of this detailed and authoritative book reads: 'the basic tenet of this field, that epigenetic processes are not fully erased during gametogenesis, runs counter to what was previously believed by epigeneticists not so very long ago'. One of the first studies showing epigenetic transmission following changes in the adult: Sharma, A. and P. Singh (2009) Detection of transgenerational spermatogenic inheritance of adult male acquired CNS gene expression characteristics using a *Drosophila* systems model. *PLoS ONE* 4:e5763. For a valuable review of this field see Burggren, W. (2016) Epigenetic inheritance and its role in evolutionary biology: re-evaluation and new perspectives. *Biology* 5:24; DOI: 10.3390/biology5020024.

26 See Rob Martienssen: 'It was actually Darwin who first realized this potential. He was a big fan of Lamarck. A lot of people don't realize that, but in *The Variation of Animals and Plants under Domestication* he wrote that if Lamarckian Inheritance – the inheritance of acquired traits – was true, then there must be some property arising in the body that could enter the germ line and change the germ line for the next generation. And Darwin called these gemmules – which is a wonderful name and we think small RNAs are very good

candidates for those gemmules.' In *The Evolution of Evolution: Darwin's Gemmule Theory Revisited,* http://bigthink.com/in-their-own-words/the-evolution-of-evolution-darwins-gemmule-theory-revisited.

27 Baldwin, James (1896) A new factor in evolution. *The American Naturalist* 30:441–451. Many biologists today regard the Baldwin effect as perfectly compatible with the Neo-Darwinian Modern Synthesis. This can be done by taking a purely gene-centric view of what is happening. That viewpoint conceals the fact that the process depends on an active choice of environment *at the level of the phenotype.* Baldwin was a psychologist and described the phenomenon as the effect of learned behaviour on evolution. The important point is that it is organisms that choose to behave in a particular way, not genes. The active role here occurs at the phenotype level. Genes then follow by the process of assimilation. The Baldwin effect is as much an assimilation of a character into the gene pool as Waddington's experiments were.

28 This is the term favoured by Patrick Bateson, who has carefully researched the literature on the 'Baldwin Effect', see Bateson, P. (2006) The adaptability driver: links between behavior and evolution. *Biological Theory* 1:342–345. He makes two points. First, that this process was first identified by Douglas Spalding in 1873, so predating Baldwin. Second, that a behavioural driver dependent on the adaptability of the phenotype could drive evolution more rapidly in a direction that would be extremely unlikely to occur by combinations of chance mutations. Both of these conclusions are convincing, so I will use 'adaptability driver' instead of 'Baldwin effect' in this book.

29 The mechanisms by which animals make such choices is now an active field of study, see Rubenstein at www.princeton.edu/eeb/people/display_person.xml?netid=dir.

30 This is a fundamental point. Just as in physics, where higher-level thermodynamics can be well ordered while random movements occur at the molecular level, so in biology apparent randomness at one level (e.g. genes as DNA sequences) does not guarantee randomness at a higher functional level.

Baldwin was a psychologist. His starting point was human cultural behaviour. It is not therefore surprising that he attributed the mechanism that carries his name to an active choice of the organisms. Wozniak, R.H. (2001) Development and synthesis: an introduction to the life and work of James Mark Baldwin. In *J.M. Baldwin, Mental Development in the Child and the Race: Methods and Processes* (Thoemmes Press, Bristol; vxxxi).

31 Odling-Smee, F.J., K.N. Laland and M.W. Feldman (2003) *Niche Construction: The Neglected Process in Evolution* (Princeton University Press, Princeton, NJ). Laland, K., J. Odling-Smee and S. Turner (2014) The role of internal and external constructive processes in evolution. *Journal of Physiology* 592:2413–2422.

32 Textbooks and articles on the Modern Synthesis usually regard the 'Baldwin effect' (adaptability driver) as compatible with Neo-Darwinism, a view that is reinforced by the fact that Baldwin was arguing against the Neo-Lamarckians of the late nineteenth century and allied himself with the first Neo-Darwinians. As the Wikipedia entry on the Baldwin effect says, 'it is generally recognised as part of the modern evolutionary synthesis'. This can be true if one really believes that it is only by chance that organisms display such adaptability. But that is to assume already what is in question regarding Neo-Darwinism, which is whether everything arises through blind chance. That seems to me to be completely contrary to Spalding's and Baldwin's intention, which was to highlight the role of active behavioural choice in organisms. This argument is at the heart of the debate over the role of purpose in living systems.

33 This process was formalised in the Wright-Fisher and Moran mathematical models of evolutionary change. Masel, J. (2011) Genetic drift. *Current Biology* 21:R837–R838; DOI: 10.1016/j.cub.2011.08.007. Moran, P.A.P. (1958) Random processes in genetics. *Mathematical Proceedings of the Cambridge Philosophical Society* 54:60–71; DOI: 10.1017/S0305004100033193. These equations are the basis of what is called the neutral theory of evolution. It can provide a random mechanism explanation for how populations can diverge over time.

34 Ticehurst, Claud (1938) A systematic review of the genus *Phylloscopus*; see http://onlinelibrary.wiley.com/doi/10.1111/j.1474-919X.1941.tb00628.x/epdf.

35 Alcaide, M., E.S.C. Scordato, T.D. Price and D.E. Irwin (2014) Genomic divergence in a ring species complex. *Nature* 511:83–85.

36 Only four examples of ring species have been described, and the evidence in two of these is not certain. These are a ring of gulls (*Larus* species) around the Arctic Circle, and a ring of salamander types around the Californian Central Valley. See http://en.wikipedia .org/wiki/Ring_species. Examples of the idea of other forms of geographical separation being a factor in speciation are much more widespread.

37 Skinner, M.K., C. Gurerrero-Bosagna, M.M. Haque, *et al.* (2014) Epigenetics and the evolution of Darwin's finches. *Genome Biology and Evolution* 6:1972–1989.

38 Skinner *et al.* (2014) Epigenetics and the evolution of Darwin's finches. *Genome Biology and Evolution* Vol. 6, Issue 8 pp. 1972–89.

39 A recent study has also shown that differences in phenotype sufficient to define different species can also be consistent with the same genotype. Mason, N.A. and S.A. Taylor (2015) Differentially expressed genes match bill morphology and plumage despite largely undifferentiated genomes in a Holarctic songbird. *Molecular Ecology*, http://onlinelibrary.wiley.com/enhanced/doi/10.1111/mec.13140: 'The Holarctic redpoll finches (Genus: *Acanthis*) provide an intriguing example of a recent phenotypically diverse lineage; traditional sequencing and genotyping methods have failed to detect any genetic differences between currently recognized species, despite marked variation in plumage and morphology within the genus.'

40 Holmes, Bob (2013) Life's purpose: can animals guide their own evolution? *New Scientist* 10 October: www.newscientist.com/article/ mg22029380.700-lifes-purpose-can-animals-guide-their-own-evolution.html. Corning, Peter (2013) Evolution 'on purpose': how behaviour has shaped the evolutionary process. *Biological Journal of the Linnean Society* 112:242–260.

41 ENCODE stands for the Encyclopedia of DNA Elements. The ENCODE Consortium (2004) The Encode Project. *Science* 306:636–640.

42 See also books by Nessa Carey: *The Epigenetics Revolution* and *Junk DNA: A Journey Through the Dark Matter of the Genome*.

43 Gould, Stephen Jay and Niles Eldredge (1977) Punctuated equilibria: the tempo and mode of evolution reconsidered. *Paleobiology* 3:115–151.

44 Margulis, L. (1970) *Origin of Eukaryotic Cells* (Yale University Press, New Haven, CT). Margulis, L. and Sagan, D. (2002) *Acquiring Genomes: A Theory of the Origins of Species* (Perseus Books Group, New York).

45 A good example is Martin Brasier's explanation of the Cambrian explosion as a kind of evolutionary avalanche, in which many factors may have built up towards shifting the evolution of life rapidly towards a new attractor. Brasier, M. (2009) *Darwin's Lost World: The Hidden History of Animal Life* (Oxford University Press, Oxford; pp. 112–118).

46 Mathematical models of genetic drift were formulated by Sewall Wright and Ronald Fisher (1931) and Patrick Moran (1958).

47 Motoo Kimura's Neutral Theory of Molecular Evolution was formulated in 1968.

48 Pigliucci, Massimo and Gerd Müller (2010) *Evolution: The Extended Synthesis* (MIT Press, Cambridge, MA). Lalan, K.N., T. Uller, M.W. Feldmen, *et al.* (2015) The extended evolutionary synthesis: its structure, assumptions and predictions. *Proceedings of the Royal Society B* 282:20151019.

49 Replicator theory is reviewed by Fernando, Chrisantha and Eörs Szathmary (2010) Chemical, neuronal and linguistic replicators. In *Evolution: The Extended Synthesis*, M. Pigliucci and G. Müller, editors (MIT Press, Cambridge, MA).

50 Noble, D. (2015) Evolution beyond Neo-Darwinism: a new conceptual framework. *Journal of Experimental Biology* 218:7–13. A short video of the transition from EES to the Integrated Synthesis can be accessed at http://jeb.biologists.org/content/suppl/2014/12/18/218.1.7.DC1/JEB106310.pdf.

51 Gerd Müller (personal communication) and I agree that the difference is largely terminology. If we define the Modern Synthesis as its original formulation to exclude the inheritance of acquired characteristics, then the EES is a replacement, not simply an extension.

52 Smith, John Maynard (1988) *Evolutionary Genetics,* second edition (New York: Oxford University Press; p. 8). On this page he also clarifies Darwin's disagreement with Lamarck, which was not about the inheritance of acquired characteristics, but rather about Lamarck's idea of an inherent drive towards complexity.

53 Third Way of Evolution website: www.thethirdwayofevolution.com. At the time of writing over 70 biologists have signed up to this website, which is rapidly growing. Some favour extension, others favour replacement, of the Neo-Darwinian Modern Synthesis.

54 The main basis of this elasticity lies in the elastic definition of a gene, discussed fully in Chapter 5. It can be made to switch between a phenotypic trait and a DNA sequence almost imperceptibly. But there has also been considerable moving of the goal posts, with even the inheritance of acquired characteristics being regarded as compatible with Neo-Darwinism on some accounts. See exchange of views in Williams, C.A. (2015) Neo-Darwinism is just fine. *Journal of Experimental Biology* 218:2658–2659. I think August Weismann would be turning in his grave!

9

The Relativity of Epistemology
The Meaning of It All

*Many humanist and religious authors . . . have drawn attention to
its [Neo-Darwinism's] damaging effects on man's spiritual life.*
(Conrad Waddington, *The Strategy of the Genes*, 1957)

As we have seen in this book, Conrad Waddington was a great and
innovative biologist, strongly opposed to Neo-Darwinism. The quotation here might nevertheless surprise you in his use of the word 'spiritual'. He specifically referred to non-religious as well as religious authors,
so he was clearly not using the word in a purely religious context. On the
contrary, he viewed ethics and other aspects of spirituality as a secular
matter in the context of evolutionary progress.[1]

From a scientific perspective I see spirituality, as I think Waddington did, as purposive creativity, which is what I will usually call it. Such
creativity arises from purposive processes in organisms. To be sure,
those processes have a material base in the molecular components. But
as you will have learnt from this book, functional purpose at higher
levels in organisms is precisely what constrains those components to
serve as its material base, while they still follow the laws of physics and
chemistry.

I don't think that scientific interpretation is widely appreciated,[2] so
this chapter begins with another story from my own experience.

At an early stage in developing the ideas that led to this book I presented some of my results using mathematical modelling of biological

systems to a meeting on the limits of reductionism in science chaired
by the great embryologist at University College London, Lewis Wolpert,
himself a leading and vocal reductionist.[3] My talk concerned the necessity for integration to complement reduction in biology. Out of the 25 or
so participants, only one openly supported my argument. I had expected
at least two others, one of them a very distinguished neuroscientist who
confided in me during the coffee break: 'I would go along with you, Denis,
except that what you are arguing would let God back in.' This was very
revealing to me, since I certainly did not think that conclusion followed
at all. Yet he clearly did, and had no doubt about it. In fact, I suspect that
for many scientists, defending reductionism, including particularly Neo-
Darwinism, was a necessity in order to counter the claims of creationist
religions or supernatural intelligent design. I think they are throwing out
the baby with the bathwater. The baby is what it is to be human, including
creative purpose; the bathwater is the murky deep water where the philosophical distinctions between science and creationist religion might be
found.

This chapter addresses the question of whether they are right to do so.
Is there a scientific perspective on what it is to be human?

Why? Questions and Goals

Perhaps the biggest 'why?' questions of all are why we are here and why
the universe is as it is. These are the questions with which this book
began.

Public debate on these questions has become so polarised that it may
seem that there are only two kinds of answer to that question. On one
side are some (but by no means all) of the world's religions, attributing
meaning to a creator that designed the universe and gave it purpose. I
wrote 'that' rather than 'who' since 'who' would restrict the idea to a
creator conceived in our image as a person. 'That' leaves the question
open to include intelligent design, whether or not one ascribes person-
hood to the process by which the 'intelligence' operates. Either way, what
are intended here are entities that are beyond what empirical science can
know about. We can view that as a definition of what is meant. Metaphysics of this kind really is 'beyond' physics and so beyond what we can
investigate.

On the other side are the reductionist interpretations of evolutionary biology as it developed during the twentieth century. Those interpretations involve the idea that every variation arises as a consequence of blind chance followed by selection amongst the variants. Blind chance is seen here as the 'real' underlying nature of the universe. That view is supported by the feeling that, at the lowest scale, the fundamental particles cannot conceivably have goals or intentions. But there is also a further assumption, which is usually unspoken. This is that if the elements cannot have goals, then nothing else in the universe can do so. There is no room in this interpretation for purpose.

By contrast with both of these extremes, 'intelligence' that we can investigate and know about is in another category which we may call 'natural purposiveness'. We have encountered it several times in this book and I will say more about it later in this chapter. Natural purposiveness is to human purposiveness much as natural selection is to artificial (meaning human) selection.

The phenomenon of having goals and purpose is called teleology. The elimination of teleology takes various forms in reductionist accounts. One of these is to represent teleological explanations as failing the test of being parsimonious. A clear message of this book is that what is parsimonious at one level is not necessarily so at other levels. When nature exhibits complexity we should acknowledge that by moving to the level that is appropriate for a successful explanation, which can then be purposive. I illustrated that idea in Chapter 3, where we saw that, at the level of molecules, it doesn't make sense to talk about heart rhythm. At the level of cells it does. And at the level of the organism it becomes a perfectly good hypothesis to identify the purpose of that rhythm.

The elimination of teleology is often presented as a virtue. In fact it is an unnecessary restriction. It expresses the metaphysical view that teleology should be avoided in scientific explanations. There is no experimental proof of this assumption. The better and practical criterion is whether theories that involve teleology satisfy the criterion of utility: do they lead to good scientific experiments and explanations? William Harvey's seventeenth-century experiments on the circulation of the blood were also experiments that revealed the purpose of the heartbeat.

That discovery has had great utility in the subsequent physiological studies of all functions of organisms with circulations. It led, for example, to the search for how arterial and venous parts of the circulation are connected and so to the discovery of capillaries. It is impossible for a physiologist to discuss the roles of oxygen, carbon dioxide and nutrients that cells require to survive without invoking the purpose of the circulation. All of the molecules involved are constrained to do what they do by the cardiovascular system as a whole. It is to this system that we can and should ascribe purpose. And, incidentally, it makes much more sense to do this than to ascribe a self to a sequence of DNA.

This is a question of what kinds of theory make sense at different levels. It is only by focusing on the molecular level that we can possibly justify the view that there is no purpose. It is an important consequence of the theory of Biological Relativity that we should ascribe functions and purposes to the level at which they make sense, which is the level at which they constrain the interactions of the system at lower levels. This constraint is also what canalises those interactions to serve the natural purposiveness of organisms. Notice also that I switched to talking about levels rather than scales. As explained in Chapter 2, levels and scales are not the same. A level, which must occur at some scale or other, is defined by the organisation that is to be found at that level.

The Third Way[4]

As we have seen in the previous chapters, there is a 'third way'. Blind chance is not the only way in which evolutionary variation occurs. Changes in DNA can be responsive to the environment within the organism itself, and to the organism's environment. These in turn have depended on the activity of previous organisms since, in response to such influences, organisms have rearranged genomes during the course of evolution. Changes in the organism in response to the environment can be inherited. DNA is not therefore 'sealed off from the outside world' nor did it 'create us body and mind'.[5] On the contrary 'Nature is even more wondrous than the architects of the Modern Synthesis thought, and involves processes we thought were impossible'.[6] We are still absorbing the immense implications of these developments in biology, and many other disciplines. Whole areas of economics, sociology and philosophy are based on interpretations of selfish gene viewpoints. No field of human endeavour will remain untouched since the implications affect even our concept of humanity.

Is it necessary that science and humanity should be in opposition? Creative purposiveness is the human characteristic that is most often 'explained away' by reductionists. Yet it is the very essence of our art, music, sport, poetry and writing, which all obviously depend on this aspect of our humanity. Science itself is also one of the greatest expressions of creativity. It is not, or should not be, a merely mechanical accumulation of data without interpretation. The activity of interpretation requires creativity. So why does the naive reductionist view have so much difficulty with it? What is the problem?

Science and Humanity

The problem can be illustrated using the story with which I opened this chapter. Can it be correct to defend reductionism, including Neo-Darwinism in particular, as a necessity in order to counter the claims of creationist religions or supernatural intelligent design? Isn't this the wrong way to defend reductionism in science? We need the reductionist approach, provided that it is complemented by its natural and necessary companion, integration. We should not let this necessity be hidden behind fears of misinterpretation by creationists.

It is the wrong way also because, in its dogmatism, it makes reductionism itself into a form of religion, in the general sense of requiring faith, not proof. There is nothing in the scientific method that justifies the view that there is only one way to study Nature. To represent science as necessarily and uniquely reductionist is therefore an article of faith. That is even more obviously true today when, as we have seen in previous chapters of this book, the Neo-Darwinist theory is clearly incomplete. It is therefore a paradox of the Neo-Darwinist and creationist debates that the dogmatic forms of both sides start from a position of faith. It is in this context that we can understand why many prominent Neo-Darwinists are also prominent atheists. That also is a statement of faith. Part of that statement of faith is that creative purpose, consciousness and intentionality are all mirages, epiphenomena without significance or effect. This leads to the idea that our actions are mechanically determined by our brains and, ultimately, by our genes. Not surprisingly, many non-scientists, and indeed many scientists also, do not think this could be anything like an adequate representation of what we know and see within and around us. The purely mechanistic view of life doesn't resonate with common sense knowledge of ourselves and of many other species.

Science and Common Sense

Many would say so much the worse for common sense. Common sense, after all, has often been wrong and it is science that has usually revealed where it goes wrong. As we saw in Chapter 1, common sense was wrong about the centre of the universe, and it was certainly surprised, even shocked, by the indeterminacy, apparent or real, of quantum mechanics. Common sense can even confuse scientists. The French philosopher August Comte famously said in 1835 that we could never know the chemistry of the stars.[7] Yet we now know so much from studying the spectroscopy of starlight that we have come to understand that we ourselves are stardust, as are all the atoms on Earth, since in the hot cauldron of stars is where atoms were formed. Comte would have needed to know about the surprising findings that led to quantum mechanics to avoid his mistake, since it is the quantal energy jumps of electrons in atoms that enable the light spectrum to reveal which atoms and molecules are present. This would have been inconceivable prior to the revolution in physics created by quantum mechanics. Comte would have thought that all starlight was much the same.

As another example, in 1895 the British scientist Lord Kelvin stated that it would be impossible to build a machine heavier than air which could fly. Just a few years later, the Wright brothers did just that with the first aeroplane in 1903. In this case, it was the application of already understood physical theory of air flow combined with clever engineering that achieved the demonstration that it was possible. Even today, many of us feel that it challenges common sense that a huge and heavy jumbo jet should be able to fly. We feel that even when we also fully understand the aerodynamics of what is happening.

It is important therefore to recognise that painstaking scientific investigation and engineering creativity have often overturned common sense. Much of our world today, with its wireless communications, internet, cloud computing, travel to planets and much else, would look impossible to our predecessors.

But we mustn't conclude from this history that science must always be right. Sometimes the clash between science and common sense is a conceptual matter that is not and cannot be settled by empirical enquiry alone. I can understand perfectly well that the displaced paving stone that made me fall and hurt myself is virtually empty at an atomic scale since its individual atoms are mostly formed of empty space. But it was nevertheless hard and solid to me when I tripped over it. The problem with this kind of clash is largely semantic. In ordinary conversation it makes perfect sense to use the words 'solid' and 'hard' in the way we do, while also understanding that, at a micro scale, there is mostly empty space. Hardness is a macro property, not applicable to the ultra-micro level. This simple example illustrates the influence of scales and levels on the way in which we explain things. Explanations at one scale are not necessarily appropriate at other scales. This is an important part of the relativity of epistemology to which I will return later in this chapter.

The clash between a mechanical view of the universe, including living systems, and a functionally purposive view is in this conceptual category. The purely mechanistic view is itself a form of metaphysics. So can we find a criterion for where 'science' might be wrong? That is surprisingly simple. Just ask whether what we think is 'science' (notice the inverted commas) is really hiding a statement of faith by making an assumption that cannot be justified by experimental evidence. In the case of the clash between mechanical and functionally purposive views of life, the unjustified assumption is that organisms are closed, determinate systems, whereas in fact they are open systems. It is that uncontroversial and well-established fact that leads to the issue between blind mechanism and meaningful function being a conceptual one. Demonstrating pure 'blind' mechanism at one level does not guarantee the absence of function at a higher level.

This is a subtle but deeply significant philosophical point, usually ignored by simplistic reductionism. We often think of factual and conceptual questions as completely separate. That is also what I was taught as a student. Science was interpreted to be simply interested in factual truth and could float free of philosophy. But factual and conceptual questions necessarily interact. If genomes really were 'sealed off from the outside world' the conceptual questions about mechanistic and functional views of organisms would be quite different. But as we have seen, the assumed 'fact' of being 'sealed off' is simply not true. We are therefore forced to consider more subtle views of organisms and of questions like 'what is life?'. I say more subtle because complexity is always more challenging, requiring more nuanced approaches, than is simplicity. The nuanced approach required is that of scale-dependent theory. As we learnt in Chapter 6, that can be quite as rigorous, quite as mathematically precise, as simple reductionism. It just requires that many more factors and forms of causation be taken into account and that we understand the principle of Biological Relativity.

Returning to the views of creationists, it doesn't get us very far to answer their theist position with what amounts to an alternative statement of faith. Some thoughtful theists have long realised this fact and, whether or not you agree with their theism, they are often better philosophers than their naive attackers. In common with many other scientists, I feel embarrassed by the lack of basic philosophical awareness in much of what is written on this matter on behalf of 'science'. Whether the authors know it or not, they are in fact speaking not on behalf of science but rather on behalf of an alternative metaphysical viewpoint, and often enough they do not appreciate the need for humility in the face of the deep uncertainties. To claim to speak to the general public with 'scientific' authority about the deepest 'why' questions with a false certainty that cannot be justified simply creates problems, it does not solve them.

Is Naive Theism the Only Alternative?

Moreover, naive theism is not the only religious answer to the biggest why question. There have been, and still are, major religions that do not require belief in a creator. Even within the Christian and Jewish traditions there are strands of thought that are closer to forms of agnosticism, and Buddhism is an obvious example amongst oriental religions.[8] There are many more. In earlier centuries the first Christian missionaries to discover these religions in East Asia needed to invent new words to convey the idea of God as creator of the universe. The word 'god' in the languages of the region would not have conveyed their message. It could just as easily have been the spirit of a stone.

These religious traditions are very different from the naive theism usually chosen as the object of attack by reductionist philosophers and scientists. Those traditions are more concerned with spirituality and practice than with belief. In the case of Buddhism, the founder even counselled against bothering much over belief when there was the much more practical question of alleviating suffering to deal with. This is the point of the parable of the poisoned arrow: rather than bother with who fired the arrow, just remove it and treat the wound.[9]

Moreover, in contrast to the difficulties in knowing what a God could or might be, there is nothing mysterious about creative purposiveness in the scientific sense in which I am using the term here. Once we recognise that life and living organisms have purpose, the very possession of goal-directed behaviour *is* to express creativity, in its various social forms.

So why should anyone think that this all 'lets God back in'? I suspect that those who think this way fear that to acknowledge creative purposiveness implies spirituality, which in turn is seen only in a religious context. To return to the quotation at the beginning of this chapter, I think that this arises because we no longer use the word 'spirituality' in the natural scientific context in which Waddington used it. Yet this is its original meaning!

The word spirit comes from the Latin word 'spiritus', meaning breath, one of the obvious processes of life in higher organisms. Many spiritual traditions involving meditation focus on the breath. Breathing is a vital process in us. Life itself is also a process, as we have seen in previous chapters. Life processes are spiritual in the relevant sense that they are not purely material. It is easy to show this. At the moment of death it is the processes of life that will cease. The molecules will remain, but no longer constrained by a living system. What therefore leaves the body at that point is not a material thing. Nor is it a ghostly 'something other'. If by 'spiritual' we mean non-material then this is its most natural scientific meaning. In this sense the mind is also spiritual. The mind is not my brain. It is not a material substance. It requires the brain, and indeed the rest of my body, to function. But to say that something is necessary is not to say that it is necessarily sufficient. It is a conceptual mistake to equate the mind with just a part of the body.[10]

These interpretations of mind–body relations enable us also to return to the difficult question of the vital force, or oriental equivalents such as the Chinese *qi*. The obvious interpretation using the conclusions of this book is that vital force simply consists in the processes of life. These processes express the creativity of living organisms. In many, or even all, organisms they also include social interactions with other organisms and with the physical environment. They express the essential openness of living organisms. This openness and the interactions with other organisms is the condition for creativity.

My reading of Lamarck's great work, *Philosophie Zoologique*, also suggests this interpretation of 'le pouvoir de la vie', usually translated into English as 'life force' and so often mistaken for a form of vitalism. On the contrary, Lamarck was firmly a materialist. Any complexity in living organisms had to originate from physical interactions. Some of his ideas on complexity would be consistent with modern systems thinking.[11]

Contextual Logic

An essential property of living processes therefore is their sensitivity to the logic of their context. Organisms with purpose act from reasons not just through causes. Reasons and causes are not the same. Of course, there are required neural, muscular, hormonal and other bodily causes of our actions, but in open systems these necessarily mesh with the processes occurring in the environment. Some of these processes are accidental, such as avalanches, storms and other events that we often call 'acts of god', and they include mechanical failure of man-made systems, and many others besides. But in the case of social interactions they are far from accidental. The reason for that is that the goal-directedness of each organism interacts and meshes with that of other organisms. The overall behaviour of the population then resembles that of games that are rule-dependent rather than that of random physical events. Games follow the contextual logic – the rules of the game and the strategies of the players – as well as the physical constraints on what is possible. This is how meaning enters into biological function.

Contextual social interaction to produce rule-dependent behaviour would be impossible if one really believes that all actions of organisms are determined solely by neural activity in the brain and elsewhere in the body since, on the contrary, some of those events are constrained by the contextual logic. This is yet another example of reductionism representing causality the wrong way round. The way out of the conceptual block of reductionism is to remember that organisms are stochastic and they are open systems. The stochasticity confers flexibility to produce a variable rather than a determinate phenotype, while it is the interactions with the environment, including other organisms, that enables each organism to display rule-guided behaviour. Living organisms exist at the interface between the stochasticity of their molecular processes and the rule (or reason) dependent interaction with their environment.

Consider what I am doing at the moment. Writing a book (or indeed any other creative endeavour) is one of the supreme examples of the consequences of organisms being open systems existing at such an interface. It is a purposive activity in response to all the reasons that lie in the context in which the book is conceived. Without knowledge of that context and its logic, analysis of the trillions of nerve discharges in my brain would get you nowhere. The only way to make sense of what I am doing at the moment is to see it as a meshing of my neural activity with that context, including the purposive behaviour of other organisms. My neurons are necessarily constrained by those meaningful interactions. Therefore the useful answer to the question of what I am doing is: 'I am writing a book.' I can then explain the reasons, and significantly those will not usually include or be restricted to physical mechanical events. An answer that details my neural activity will be of interest to a neuroscientist, but it will not and cannot answer the question. Nor would it naturally lead to those reasons. To say therefore that those neural events are all that is happening is to ignore precisely what is purposive about such activity.

The reductionist reply to this kind of argument is to quote the important neurophysiological experiments of the American scientist Benjamin Libet. In 1983 he showed that it was possible to record activity in the brain when a human subject was executing a voluntary action like pressing a button. The important observation was that such neural activity can occur before the subject notices that he has decided to make the movement.[12] The reductionist interpretation of the experiment is that the appearance of freely choosing to make the action is just that, an appearance not a reality, since the brain 'acted' before the subject had the awareness of doing so. If one considers the brain to be a purely mechanical closed system, this must surely mean that the subject was forced by his brain to make the choice.

There are in fact many possible interpretations of the experiment, which turns out not to be as clear-cut as originally thought. The one that is relevant to this book is that we must take account of the social context in which the subject is acting.[13] He is told that he should choose when he wishes to make his action. Is it surprising that this should result in build-up towards doing so? This is the logical context in which the expectations of needing to act interact with the neural states of the subject's brain. Those interactions are going on all the time while writing this book. I experience various kinds of impulse, to stop, to return, or not, to doing it. I have no doubt that I am often on auto-pilot when pacing the garden, cooking a supper, playing the guitar, reading a new article. Preparatory neural activity must be occurring all the time. Yet my intention to write the book is always there. Intentions that really matter in our lives are like that. They are not discrete single events in the way in which the Libet experiment supposes. Moreover, we can be certain that the words that eventually appear on the screen of my computer are not in themselves anything other than the product of the meshing of my neural and bodily activity with the complete contextual logic of my interactions with all of those activities and the environmental influences on them, the latest of which was re-reading Libet's 1983 and subsequent papers. As I read those papers I also knew that he intended to write them and that they also arose from just such a process of interactions with the contextual logic of his situation.

To explain this kind of meshed interaction between internal and external factors in a way that may help the reader, I will use an extension of the diagram of Conrad Waddington that we used in Chapter 6. Instead of genes at the lower level, we will have neuronal activity; instead of the gene–phenotype networks we will have neural and hormonal networks. Crucially, to those networks we now add network links to the environment and to other organisms. These links were always implicit in Waddington's idea, but we now make them explicit. As in his original diagram, those lines of interactions are the means by which the networks constrain the landscape, which now becomes a landscape of possible behaviour rather than a landscape of development. This version of his diagram is also valid for Waddington's original application of the diagram to genes and their interactions through networks.

The landscape in this form of the diagram is shaped by the networks of interactions in the environment as well as by networks within the organism. Some of these interactions, including particularly those with other organisms, will exhibit purpose since all organisms do that to varying degrees. The mesh between those interactions and the goal-directed response within the organism itself is what leads to rule-guided

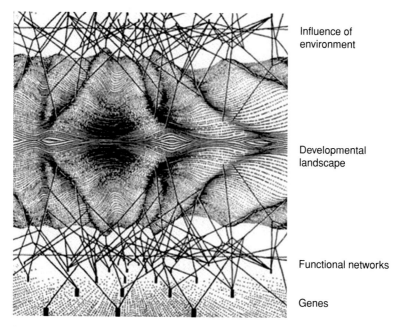

Influence of environment

Developmental landscape

Functional networks

Genes

Figure 9.1 Extension of Waddington's diagram to represent the open nature of organisms. The bottom half of the illustration is the same as his original diagram, already shown in Chapter 6 (from Waddington, 1957). The upper half represents environmental influences. Both influence the phenotype landscape. Unlike the networks in the bottom half, which link to specific genes, the network of environmental influences opens up to an indefinitely large set of interactions, which can only be hinted at in the diagram by the lines disappearing at the top. Nothing in the environmental networks corresponds to specific 'genes'. A similar extension of Waddington's diagram can be found in Bateson (2014).[14]

behaviour, which we can view as a social interactions game. This is the most dramatic of the forms of downward or contextual causation, because many of the outcomes will be impossible to predict from knowledge of the single organism. It is dramatic also because rule-guided behaviour is precisely behaviour for which there are reasons, not just physical causes. Drama, of course, requires meaning. Meaning is a social phenomenon (Figure 9.1).

This may seem strange to someone who is locked-in to a purely mechanistic view of living systems. How, they will ask, can a sequence of 'blind' physical events come to embody a process that is driven by reasons, that follows rules, including moral rules?

Yet it should not seem strange. The readers of this book almost certainly possess machines that do this. Computers are, at the physical level, deterministic machines. Yet, when they are following a logical program they follow rules, i.e. the contextual logic formed by the program. Like living organisms, modern computers are open systems that interact with us and with their environment.

But wait a minute, you may say. That is so only because a human wrote the program, not because the logical behaviour emerged from a purely physical system. Precisely so. The openness of the computer as a machine is exactly what enables its otherwise purely mechanical functions to be constrained by the contextual logic derived from its interaction with humans. Even if you wish to maintain that a human being is also a physical automaton, you would have to grant that its openness also leads to its 'programming' by the social interactions into which it enters. Together with stochasticity that is what ensures that we end up not being purely mechanical machines. Otherwise, why would I be writing this book? I do so because I have a reason to do so; to convince you of the theory of Biological Relativity and its implications. There is a vast contextual framework to writing this book. You, the reader, will readily understand that. You do not need to examine the neurons in my brain to do so.

The equivalent to the human writing a program for a computer is the evolutionary process. Goal-directed behaviour has evolved. Creative evolution 'wrote' the 'programs' for the logic of living systems, but it did not do so solely in the genome. Nor did it isolate the genome from itself and its environment.

I put 'wrote' and 'programs' in inverted commas to warn us that we then no longer need the program metaphor. The program becomes the process itself, not a separate set of instructions.

Selfish Genes and Altruism

A serious challenge to the Neo-Darwinist Modern Synthesis is how to account for genuinely altruistic behaviour. A very limited form of altruism can be derived from the idea that individual organisms might preserve their genes through acting altruistically towards other members of their species who share those genes. This is the approach that gave rise to kin-selection theory since the organisms that would be most likely to share the same genes (strictly speaking, the same gene variants) will be near relatives.[15]

It is of course true that close kin often act altruistically towards each other. But there are nevertheless two obvious problems with this explanation. The first is that it clearly doesn't explain altruistic behaviour towards other organisms that are not closely related. The second and even more serious problem is that it conflates the concepts of selfishness and altruism at different levels. This is a mistake since selfish or altruistic behaviour at one level does not automatically guarantee such behaviour at another level. A selfish individual nevertheless relies on extensive co-operation between its genes (and its proteins, lipids, etc.) in order for it to be even possible that he or she might be selfish as a whole individual organism. Whether or not our genes can be represented as selfish, it does not follow that 'we are born selfish'.[16]

It is instructive to compare the difficulty Neo-Darwinism has to explain genuine altruism with the problem that faced Descartes three centuries ago. Like the Neo-Darwinists, Descartes also imagined that it would be possible to 'deduce the whole form and configuration' of an organism from knowledge of what was in the semen, as we saw in Chapter 6. Descartes' way out of the dilemma this causes in trying to explain human behaviour was to suppose that the mind interacts with the body (through the pineal gland) to enable the mind to control the body. Curiously, this is exactly what some Neo-Darwinists take as their way out of the dilemma. Consider this text from *The Selfish Gene*:

> Let us try to teach generosity and altruism, because we are born selfish. Let us understand what our own selfish genes are up to, because we may then at least have the chance to upset their designs, something that no other species has ever aspired to.[17]

The comparison with Descartes' position is appropriate because Dawkins not only supposes that we can 'upset their designs' but also that this is unique to our species, as did Descartes. This is therefore Cartesian dualism, though I imagine that it is unlikely that Neo-Darwinians would wish to acknowledge that. The mysterious nature of the intervention is also a parallel. There is no explanation of the process(es) by which the 'upsetting' may occur.

As we saw in Chapter 6, there is no mystery here. Processes at higher levels in organisms naturally constrain the components at lower levels through determining the boundary conditions within which those components interact. We don't require overt or covert recourse to Cartesian dualism. In open systems, it necessarily follows that analysis of behaviour at a lower level will be incomplete in precisely the way in which differential equations of motion are incomplete until the conditions necessary to integrate the equations have been incorporated. This is a central aspect of the theory of Biological Relativity.

Relativity of Epistemology

In Chapter 2 we saw the immense differences in scale that are relevant to living organisms. Just to repeat the main point, a proton (a hydrogen ion) views a cell much as we view the universe. And that is also how a single cell views us. Universes within universes within universes – and beyond what we can see and know about, who knows?

'Know' is the right word here. We have no idea what, if anything, could lie beyond what we see and observe. That should inspire humility. So, the first step towards appreciating the relativity of epistemology is to acknowledge that obvious fact. Note also that the Latin for 'to know', *scire*, is also the basis of the word 'science', *scientia*. Necessarily, science is concerned with what we can know.

Cosmologists have shown great insight in extrapolating from what we observe now to how it may have begun and how it then developed. The big bang theory is an immensely audacious achievement, making perhaps the best sense we can of what is known at the present state of that knowledge. We can even detect the radiation that may represent what remains from the earliest events, calculate when stars began to form ('let there be light'!), and how old the most distant visible galaxies must be.[18]

So, I would like now to return to where this book began. What did our ancestors think when they observed the extraordinary firmament above us on a clear, moonless night? What now would be our answer to their big 'why' question?

They would be astonished to discover that neither the Earth nor the sun is the centre of the universe, that there isn't a centre anyway, and that the universe is vastly bigger and older than they could possibly have imagined. After asking how we now know all of that and explaining the steps by which we reached these conclusions, I can also imagine them scratching their heads. Yes, they would say, that's all amazing and very interesting. But that still doesn't answer our big 'why' question.

If they probed even deeper into what we don't know they would find further big surprises. In fact the questions could be quite embarrassing. Let's imagine the conversation:

'You tell us the universe began as infinitesimally small over 13 billion years ago, but why did it do that? Why at that time rather than any other? You tell us that the frame of reference, space-time, is itself shaped by what exists and that what exists could itself be structurings of space-time. But you also say that you don't know how to construct a theory that unifies the major theories of quantum mechanics and relativity. Isn't that rather odd?'[19]

If they looked carefully at the equations themselves they would find an even stranger fact. The cosmological constants need to be adjusted to an extraordinary degree of accuracy to make the existing universe possible. Doesn't that imply that somewhere the theories have gone wrong?[20]

How should we respond to all of that? We would surely have to say that we really don't know. But we could then go on to explain that all knowledge is relative and that it may be asking the wrong question to require an ultimate answer. A theory of everything may be a goal towards which one can progressively move, but, like trying to accelerate an object towards the speed of light, it may be impossible to say, finally, that we got there. Our ancestors would therefore be right to be astonished at what we have discovered, but they would also be right if their reaction was even deeper puzzlement.

My, admittedly tentative, conclusion is to say that this may be a bound-ary across which we cannot go. It could itself be a manifestation of the relativity of epistemology. From a practical viewpoint, though, does it matter? If, as I suggest, we can account for goal-directedness in evolu-tion without having to suppose that this requires that the universe itself should have an ultimate goal, isn't that sufficient? We return to the insight of the parable of the poisoned arrow. Why be too concerned to know who shot the arrow when the urgent action is to remove it and treat the injury?

It was another of the great philosophers, Kant, who understood this relativity of epistemology from a different perspective. He showed that we always need a framework within which to interpret the world, but that the framework itself may not be derivable from what we already know. That is why all science requires a metaphysics if it is not to be mere cataloguing. That metaphysics may not be derivable from purely empirical observations.

To use the terminology of Karl Popper, we can make conjectures that we may be able to refute, but we can never be absolutely certain that they are true, nor that there will not be a better conjecture.

Ultimate Purpose?

There may or may not be an ultimate purpose, whatever that may mean. I doubt whether empirical science can answer that question. But it certainly can use the approach of Biological Relativity and the relativity of epistemology to explain why organisms have genuine purposes and why creative purposiveness as defined and used in this book is a real phenomenon that arises naturally from the social interactions of organisms and which is therefore within the range of knowledge of science.

To return to the very beginning of the book, we have seen much further than our ancestors could when they looked up at the firmament. We have looked back in time across aeons that they would have found unimaginable. We have also observed inside ourselves and other living organisms and have done so right down to the molecular level.

But, if the history of science tells us anything about the big why questions, there can't be much doubt that future centuries will see discoveries beyond what we can imagine today. I suggest that there will always be a relativistic 'beyond' – beyond what we can know.

Notes

1 See Bowler, Peter (2001) *Reconciling Science and Religion* (University of Chicago Press, Chicago, IL; p. 80): 'Waddington had no interest in encouraging scientists to revive an interest in religion.' This is also clear from Waddington's 1942 book *Science and Ethics*. His concern

was for scientists to be involved in the social process. To quote Bowler again, 'his approach to ethics was more in line with Huxley's vision of humanity continuing the course of evolutionary progress by more efficient means'.

2 An interesting exception can be found in Capra, F. and P. Luisi (2014) *The Systems View of Life* (Cambridge University Press, Cambridge; chapter 13). Their approach to science and spirituality is very similar to mine. See particularly pp. 282–285, Science versus religion: 'a dialogue of the deaf?'.

3 Novartis Foundation (1998) *The Limits of Reductionism in Biology* (Wiley, London).

4 See the Third Way website at www.thethirdwayofevolution.com/people.

5 Quotations from Dawkins, R. (1976) *The Selfish Gene* (Oxford University Press, Oxford).

6 From Noble, D., E. Jablonka, M.J. Joyner, G.B. Müller and S.W. Omholt (2014) Evolution evolves: physiology returns to centre stage. *Journal of Physiology* 592:2237–2244.

7 Comte, Auguste (1842) *The Positive Philosophy*, book II, chapter 1.

8 The thirteenth-/fourteenth-century theologian Meister Eckhart is a good example in the Christian tradition. He referred to God as 'no-thing'. His German text refers to absolute detachment (*Abegescheidenheit*), which he clarifies as 'pure nothing'. This is not its modern meaning, which is 'departed' in the sense of 'deceased'. In the Jewish tradition, see Exodus 33:20: 'You cannot see My face, for no man can see Me and live!', which is echoed in several New Testament verses. Buddhism is well-known to be an essentially agnostic religion.

9 The Buddha is said to have declined to answer metaphysical questions. The parable of the poisoned arrow comes from the earliest teachings of the Buddha. It expresses the focus on what to do, rather than on what to believe.

10 Bennett, M.R. and P.M.S. Hacker (2003) *Philosophical Foundations of Neuroscience* (Blackwell, New York).

11 Even a casual reading of Lamarck's work makes it clear that he was not a vitalist. The French historian of genetics André Pichot wrote an excellent introduction to the 1994 reprint of *Philosophie Zoologique*, in which he says (my translation) 'Lamarck's claim that . . . there is a

radical difference between living beings and inanimate objects might lead people to think that he was a vitalist. But he is not. On the contrary, his biology is a mechanistic reply to the physiological vitalism of Bichat, which was then the dominant theory', in (1994) *Philosophie Zoologique* (Flammarrion, Paris; p. 20).

12 Libet, Benjamin, Curtis A. Gleason, Elwood W. Wright and Dennis K. Pearl (1983) Time of conscious intention to act in relation to onset of cerebral activity (readiness-potential): the unconscious initiation of a freely voluntary act. *Brain* 106:623–642; DOI: 10.1093/brain/106.3.623.

13 Some of the arguments in Julian Baggini's book *Freedom Regained: The Possibility of Freewill* also emphasise the roles of reasons and contexts. This chapter was written before that book was published, but I would like to acknowledge the important similarities. Baggini, Julian (2015) *Freedom Regained: The Possibility of Freewill* (Granta Publications, London).

14 Bateson, P. (2014) Evolution, epigenetics and cooperation. *Journal of Biosciences* 39;2:191–200; DOI: 10.1007/s12038-013-9342-7.

15 This theory was first proposed by Fisher and Haldane during the 1930s, and later developed mathematically by Hamilton and Price. The alternative approach, already referred to in Chapter 7, is to develop the equations as part of a multi-level selection model, as done by Nowak, Tarnita and Wilson. From such a viewpoint the gene-centric explanation of altruism is incomplete. Needless to say, the Nowak *et al.* model is more compatible with the theory of biological relativity.

16 Note also the religious overtones of that famous phrase: a theory of original sin?

17 Dawkins, R. (1976) *The Selfish Gene* (Oxford University Press, Oxford; chapter 1).

18 Just possibly, there might also be evidence for what may have led to the big bang in the pattern of the background cosmic radiation. A theory proposed by Roger Penrose leads to this idea. The singularity in Einstein's field equations at the big bang would then be only an apparent singularity. See Gurzadyan, V.G. and R. Penrose (2010) Concentric circles in WMAP data may provide evidence of violent pre-Big-Bang activity. *Cosmology and Nongalactic Astrophysics*: http://arxiv.org/abs/1011.3706.

19 For a recent account of this problem, see Johnson, J.R. (2015) Discovering Nature's hidden relationships, an unattainable goal? *Physics International* 6:3–10; DOI: 10.3844/pisp.2015.3.10. 'The Standard Model of Particle Physics (SMPP) and the Standard Model of Cosmology (SMC) have no elements in common, they are completely disjoint.'

20 The astronomer Martin Rees wrote a brilliant account of this problem in his (1999) book *Just Six Numbers: The Deep Forces that Shape the Universe* (Weidenfeld & Nicolson, London). Nearly two decades later we are no nearer any explanation of the remarkable sensitivity of models of the universe to the values of the constants required to enable the mathematical equations to predict a universe in which life is possible. Rees maintains a nice balance between enthusiasm for what scientific investigation can and will reveal, and scepticism about exaggerated claims for science: 'This perspective should caution us against scientific triumphalism – against exaggerating how much we'll ever really understand of nature's intricacies.'

10

Postscript

Why did the narrowly reductionist Neo-Darwinist view dominate biology for a whole century?

There are many possible explanations, which will doubtless interest historians and sociologists in the future. I can see two factors that have clearly influenced me and can explain why so many biologists did not question it. Recall, as explained in the introduction to Chapter 6, that I began my scientific career firmly in the reductionist camp. I know why it is attractive.

First, nineteenth- and early twentieth-century science had to distance itself from all manner of fantastic, not to say ghostly, views of the world. In doing so it retreated too far into what became a central citadel to be defended vigorously at all costs, almost afraid to admit anything strange and inexplicable in the universe. Physics peeled away from the entirety of that view when quantum mechanics and Einstein's relativity theories appeared. But all of that passed by with little notice in biology. The ultra-micro and the ultra-macro scales seemed of little relevance to biology. As this book has shown you, the general principle of relativity does have important consequences for biology. It changes our view of causation.

Second, that citadel came to be surrounded by a simple protective narrative that made it seem obvious that it was the repository of the only true faith. Once the Weismann Barrier became accepted (and then seemed to be strengthened by the Central Dogma) it appeared to be true that, even if organisms are open systems, their genetic material is not. In such a world the only way in which evolutionary change could occur would be by random (non-functional) variations in the genetic material.

The narrative was completed by noting that evolution by natural selection is then the only possible way. As popularisers often put it: 'it's obvious, so why do you need to drag in X?' Where, for X, read all the processes that were dismissed and all the people who were ostracised.

Can Biological Relativity be given a similarly simple and convincing narrative? Complexity often loses out in popular debate because its more nuanced approach is perceived to be not so easy to articulate. Appealing to the gallery with simple slogans does not seem possible.

But, I think that, in summing up this book, it is perfectly possible to formulate an alternative that is just as clear and just as obvious. It goes like this:

- Organisms, including their genetic material, are necessarily open systems.
- Open systems are necessarily influenced by processes at larger scales. That is what we mean by an open system.
- Meaning and function are natural features of larger-scale phenomena in biology, not of individual molecules.
- The reason is that physico-chemical processes at smaller scales are necessarily constrained by higher scales. Even a molecular determinist has to admit this.
- What is ordered and functional at higher scales can appear stochastic (random) and non-functional at lower scales.
- Lack of purpose at the molecular scale does not therefore entail lack of purpose at other scales.

All pretty clear, and after reading this book, I hope they will appear fairly obvious to the reader.

That's it in a nutshell. This is your new set of glasses.

Glossary[1]

Lamarckism
In this book I use the simplest and clearest definition of Lamarckism, which is the inheritance of acquired characters. This definition is readily testable. The idea of the inheritance of acquired characters, which is usually seen as the hallmark of Lamarckism, was commonly accepted in Lamarck's day, and Lamarck did not claim credit for it. He did claim credit, however, for the idea that the inheritance of acquired characters could lead to the transformation of species. During the twentieth century, Lamarckism became associated mainly with the view that the inheritance of acquired characters played a significant role in evolution, and was seen as an alternative to Neo-Darwinism and to the Modern Synthesis.

If Lamarckism is to mean anything, then defining it as the antithesis of Neo-Darwinism (Modern Synthesis) is appropriate. The counter-argument to this definition is that it is not what Lamarck meant. Moreover, Lamarck could not have anticipated the various epigenetic and other mechanisms now discovered by which the inheritance of acquired characteristics can occur, any more than Darwin could have anticipated the re-discovery of Mendel's genetic experiments and the formulation of Neo-Darwinism. Lamarck was not a Lamarckian in the modern sense. Nor was Darwin a Neo-Darwinist. It is important to recognise the debate over the definition of Lamarckism because the founding father of epigenetics, Conrad Waddington, was not a Lamarckian in the sense that became prevalent since Weismann in the late nineteenth century – Lamarckism became identified then with a transmission of information from the soma to germ line, not with the general notion of acquired characters that can happen through Mendelian mechanisms. A large spectrum of evolutionary doctrines became associated with Lamarck's evolutionary theory that focus on the evolutionary role of individual acquired (e.g. learnt or induced) variations that emerge during the life of an organism in response to environmental changes. Concepts such as the power of life, or the tendency for inexorable, predetermined growth, also became associated with Lamarckism, and were appropriated by philosophical, teleological and at times theologically oriented views of evolution.[2]

Darwinism

Darwin's theory of evolution by descent with modification is usually limited to that part of the theory dealing with the mechanism of evolutionary change, i.e. natural selection of heritable variations. Natural selection, usually through competitive interactions, also features in Neo-Darwinism, which elevates it to being both necessary and sufficient to explain evolution. Darwin was far more cautious. He also acknowledged the inheritance of acquired characteristics, on which he even formulated a mechanism (his theory of pangenesis), and emphasised in *The Origin of Species* that natural selection was not the only mechanism of evolution. I see the pluralist view of evolution that has been re-emerging in recent years as, in some senses, a return towards Darwin's more cautious position and away from strict and dogmatic Neo-Darwinism. But we should not read too much into this. Darwin could not have anticipated the developments in genetics, molecular biology and epigenetics that have undermined strict Neo-Darwinism, any more than he could have anticipated Neo-Darwinism itself.

Neo-Darwinism and the Modern Synthesis

The term Neo-Darwinism was originally used by Romanes in 1896 to describe Weismann's and Wallace's version of Darwinism, which specifically excluded the inheritance of acquired characters. It is now commonly used to describe the Modern Synthesis version of Darwinism, a wide-ranging consensus about the nature and dynamics of evolutionary change that emerged amongst biologists between the 1920s and 1950s, and was based on Neo-Darwinian assumptions, including gradualism, and population genetics based on the assumption that heritable variations are random (functionally blind). However, genetic change is not always random (see the glossary item on **Randomness**), is not always through gradual accumulation of small mutations, and the exclusion of the inheritance of acquired characteristics is clearly wrong. This leaves a puzzle. Many biologists who acknowledge these points still regard themselves as adhering to the Neo-Darwinist Modern Synthesis, as though it can be effortlessly extended to include almost anything that has been shown to occur in nature. In my experience this is in part because many biologists do not recall just how restrictive the original formulation of Neo-Darwinism is. And the language of Neo-Darwinism is strongly compelling until one deconstructs it (see Chapter 5 of this book). I believe it is more honest to admit that the original formulation is no longer useful in a context in which so many additional mechanisms of evolutionary change have been found.[3]

Randomness
Defining what is meant by 'random' is itself a major field of enquiry in mathematics, computation and in science generally. The question of whether there are truly random events in the universe is a vexed one, lying at the heart of theories of quantum mechanics (see Chapter 1). Probably, we will never know, perhaps cannot know in principle, the answer to that kind of question. But it warns us that defining randomness is not easy.

The best way to sidestep the deeper problems is to ask the question 'random with respect to what?'. In evolutionary theory that makes the problem much simpler. Both Neo-Darwinists and their opponents can then agree that what is really meant is 'random with respect to physiological (phenotypic) function'. That is so because one of the central tenets of Neo-Darwinism is the exclusion of any form of Lamarckism, the idea that function, or functional improvement, can influence inheritance. By contrast, a Lamarckian must maintain that at least some changes are not completely random with respect to function.

I approach this question in stages. The first stage is to establish that genomic change is not random with respect to location in the genome. The reason for asking that question first is that without establishing that there are preferred locations of change, the argument for any kind of functionally relevant change in the genome cannot even get off the ground. The only way in which such a change can occur is through influencing the physical and chemical properties of the genetic material. Preferred locations of change are therefore a pre-condition for functional change to be possible. If all locations in the genome were equipotent for changes there would be no possibility for functionally relevant change. This approach is taken further in the item on **Function,** and in Chapter 7.

Function[4]

Function in physiology or evolutionary biology is usually defined as the consistent causal role that a part plays within an encompassing human-designed or natural-selection-designed system, a role that usually contributes to the goal-oriented behaviour of this system. Since, from an evolutionary perspective, the most general goal-directed behaviour of the encompassing system is the reproductive success of the organism, the notions of function and adaptive evolution through natural selection are closely related.[5]

Genes

There is no universally correct or accepted definition of 'gene'. Even defining it as a DNA sequence coding for a protein (which is the usual modern definition) creates ambiguities since many DNA sequences that do not code for proteins also serve functional roles, and many different proteins can be coded by the same DNA sequences, depending on how splicing and recombination occur. Some of the definitions of a gene are even incompatible in their consequences for evolutionary biology. The idea of Biological Relativity sidesteps many of these problems by making it clear that there is no privileged level of causation and that the nature of that causation with regard to genes naturally depends on the definition. The reader is therefore best referred to Chapter 5 of this book and in particular to Figure 5.3 and the text discussing it.

Epigenetics

'Epi' means 'above', so epigenetics is 'above' genetics. The concept was first introduced in the later 1930s by Conrad Waddington to refer to 'the branch of biology which studies the causal interactions between genes and their products which bring the phenotype into being'.[6] The term now refers to the study of developmental processes in prokaryotes and eukaryotes that lead to persistent, self-maintaining changes in the states of organisms, the components of organisms or lineages of organisms. Epigenetic mechanisms underlie developmental plasticity and canalisation. At the cellular level, these mechanisms (e.g. DNA methylation of histone modifications) establish and maintain the changes that occur during cell determination and differentiation in both non-dividing cells, such as brain cells, and dividing cells, such as stem cells. At higher levels of biological organisation, epigenetic mechanisms underlie self-sustaining interactions between groups of cells, and between the organism and its environment. Genetics in the sense of DNA inheritance is under extensive control by many epigenetic mechanisms. Epigenetics can therefore be viewed as the way in which organisms control and use their genomes.

Natural Genetic Engineering and Mobile Genetic Elements ('Jumping Genes')

These are the genomic modifications that are the result of a dedicated machinery that alters the structure of the genome during sexual reproduction (e.g. recombination), and following various genomic stresses (e.g. repair mechanisms and transposition). Barbara McClintock, working on maize in the 1940s, was the first to show that genetic material can move around the genome, even from one chromosome to another, and suggested that stress increases the rate of transposition, but she was only awarded the Nobel Prize many decades later, in 1983, for her discovery. We now know about many additional molecular mechanisms by which transposition can be achieved. Reverse transcription from RNA to DNA enables mobile RNAs to be inserted as DNA into the genome. There are also cut-and-paste mechanisms that do not require an RNA intermediate.[7]

Reductionism

Reductionism is the view that from knowledge of the structure and reactions of the individual components of a system, and using that knowledge alone, it is possible to understand and completely predict the behaviour of the whole. In biology the components usually referred to in reductionist views are molecular.

Integrationism and Emergence

Integrationism is the opposite view that properties of the whole emerge as the interactions both amongst the components and through interactions with the environment of the system. Knowledge of the components alone is therefore not sufficient.[8]

Holism

Holism is a version of integrationism that emphasises the behaviour of the whole.

Notes

1 I thank the authors of the books and articles listed below for the material for this glossary.

2 Jablonka, Eva and M. J. Lamb (1995) *Epigenetic Inheritance and Evolution: The Lamarckian Dimension* (Oxford University Press, Oxford; pp. 30–54). Jablonka, E. and M. J. Lamb (2014) *Evolution in Four Dimensions* (MIT Press, Cambridge, MA). Snait, S.B. and E. Jablonka (2011) *Transformations of Lamarckism* (MIT Press, Cambridge, MA).

3 Noble, D. (2015) Evolution beyond Neo-Darwinism: a new conceptual framework. *Journal of Experimental Biology* 218:7–13.

4 Roux, E. (2014) The concept of function in modern physiology. *Journal of Physiology* 592:2245–2249.

5 Jablonka, E. (2002) Information: its interpretation, its inheritance, its sharing. *Philosophy of Science* 69:578–605.

6 Noble, D. (2015) Conrad Waddington and the origin of epigenetics. *Journal of Experimental Biology* 218:816–818.

7 Shapiro, J. (2011) *Evolution: A View from the 21st Century* (FT Press Science, Upper Saddle River, NJ).

8 The concept of emergence is hotly debated in philosophy. See Bedau, M.A. and P. Humphreys (2008) *Emergence: Contemporary Readings in Philosophy and Science* (MIT Press, Cambridge, MA).

Index

Locators followed by 'n' refer to notes. Locators followed by 'g' refer to the glossary.